Biodiversity 101

Melina F. Laverty, Eleanor J. Sterling, Amelia Chiles, and Georgina Cullman

Science 101

GREENWOOD PRESS
Westport, Connecticut • London

Library of Congress Cataloging-in-Publication Data

Biodiversity 101 / Melina Laverty ... [et al.].
    p. cm. — (Science 101, ISSN 1931–3950)
  Includes bibliographical references and index.
  ISBN: 978-0-313-34120-5 (alk. paper)
  1. Biodiversity—Popular works.   I. Laverty, Melina F., 1970–
II. Title: Biodiversity one hundred one.
  QH541.15.B56B5643   2008
  333.95—dc22   2008028518

British Library Cataloguing in Publication Data is available.

Library of Congress Catalog Card Number: 2008028518
ISBN: 978-0-313-34120-5
ISSN: 1931-3950

First published in 2008

Greenwood Press, 88 Post Road West, Westport, CT 06881
An imprint of Greenwood Publishing Group, Inc.
www.greenwood.com

Printed in the United States of America

The paper used in this book complies with the
Permanent Paper Standard issued by the National
Information Standards Organization (Z39.48–1984).

10 9 8 7 6 5 4 3 2 1

CONTENTS

SERIES FOREWORD

What should you know about science? Because science is so central to life in the 21st century, science educators believe that it is essential that *everyone* understand the basic foundations of the most vital and far-reaching scientific disciplines. *Science 101* helps you reach that goal—this series provides readers of all abilities with an accessible summary of the ideas, people, and impacts of major fields of scientific research. The volumes in the series provide readers—whether students new to the science or just interested members of the lay public—with the essentials of a science using a minimum of jargon and mathematics. In each volume, more complicated ideas build upon simpler ones, and concepts are discussed in short, concise segments that make them more easily understood. In addition, each volume provides an easy-to-use glossary and an annotated bibliography of the most useful and accessible print and electronic resources that are currently available.

PREFACE

Biological diversity is the key to the maintenance of the world as we know it.
(E.O. Wilson, 1999)

The incredible diversity of life on earth today has evolved over the last 3.5 million years. It has only been in the last 300 years that we have begun to describe the species on the planet, and even now many species remain unknown. Just as we begin to uncover and understand the earth's ecosystems, species are disappearing. We have entered what many scientists consider to be the Sixth Major Extinction. In the last mass extinction, 65 million years ago, about two-thirds of species living at the time went extinct, most famously the dinosaurs. According to the World Conservation Union (IUCN), 16,306 species are now endangered with extinction. We have already witnessed the disappearance of much of the world's grasslands and dry forests, ecosystems well-suited for agriculture. Though the world has experienced mass extinctions in the past, this one is different. It is driven by humans, their growing populations, and their insatiable demand for the earth's resources.

Human activity reaches to the far corners of the earth; pollution impacts even the world's remote places. There are now 6.6 billion people on the planet; yet many of our closest relatives—the Great Apes—are endangered, some critically endangered (IUCN 2008). In the last 50 years, populations of bonobos, chimpanzees, gorillas, and orangutans have declined by 50 percent, for some subspecies declines have been even greater (IUCN 2008; Cincotta and Engelman 2000). While human populations are in the billions, the great apes only number in the thousands. For example, there are only an estimated 172,700–299,700 chimpanzees (*Pan troglodytes*)—the great ape with the largest population (IUCN 2008).

Biodiversity 101 explores: What biodiversity is? Where biodiversity is found? Why it matters? How it is currently threatened? And How people can help stop the loss of biodiversity?

Despite the current biodiversity crisis there are signs of hope. Species on the brink of extinction have been successfully reintroduced and some ecosystems, such as the Amazon and boreal forest, are relatively intact. There is also a growing interest and opportunity for people to find harmonious and sustainable ways of living with nature. We hope this book will inspire people to understand and connect with the natural world, and to take action to help protect it. All life—including humans— depends on a healthy earth for survival.

ACKNOWLEDGMENTS

We would like to thank the many people who have contributed to this book. We would particularly like to thank our editor, Kevin Downing, as well as Kevin Frey, Ryan Luci, Andrew Perkins, and our colleagues at the Center for Biodiversity and Conservation for their thoughtful contributions, comments, and suggestions on the manuscript. We would also like to thank Ho Ling Poon for her efforts in locating photographs.

1

WHAT IS BIODIVERSITY?

Brilliant scarlet macaws, scampering gray squirrels, towering giant sequoia trees, your morning cup of coffee, industrious honey bees—mundane and spectacular, economically fundamental and simply arcane—all of these are examples of biodiversity, the dazzling variety of life on earth.

Few people are familiar with the word "biodiversity," yet everyone is intimately connected with biodiversity in their daily life. Humanity's fundamental reliance on and connection with natural systems creates an imperative to understand and protect biodiversity.

To conserve biodiversity, we need to understand what biodiversity is, determine where it occurs, identify strategies to conserve it, and track over time whether these strategies are working. The first of these items, knowing what biodiversity is, and therefore what to conserve, is complicated by the remarkable diversity of living things themselves.

Life on earth today is the product of about 3.5 billion years of evolution. An estimated 1.75 million species have been discovered and described, but this only represents a fraction of all the species on earth. Estimates of how many species remain to be discovered range from 3.6 million to 117.7 million, with 13 to 20 million being the most frequently agreed upon by scientists. One reason the exact number of species is still unclear is thats new species are continually being described. Some of this uncertainty is also due to the increased information available to scientists since the advent of genetic analyses and because the definition of what constitutes an individual species changes.

This chapter introduces the basics of biodiversity: what biodiversity is and how to measure it. It also explores the evolution of biodiversity over time: how many species there are today, how many have disappeared, and the "Sixth" extinction—the current rapid loss of biodiversity around the world.

DEFINITION OF BIODIVERSITY

Scientists first coined the term biodiversity, a contraction of the phrase, "biological diversity," in the 1980s. Most simply, biodiversity describes the entire variety of life on earth. It can also be defined more broadly incorporating not only living organisms, but also their complex interactions with one another and with the nonliving aspects of their environment. *Biodiversity* is defined as:

> The variety of life on Earth at all its levels, from genes to ecosystems, and the ecological and evolutionary processes that sustain it.

The Biodiversity Hierarchy

To understand and appreciate the full variety of life encapsulated by the term "biodiversity," scientists describe it based on a nested hierarchy, beginning at the subcellular scale and ending at the continental level (see Figure 1.1). The smallest level of this hierarchy refers to the diversity of genes that can be found in individual cells. *Genetic* diversity is sometimes called the "fundamental currency of diversity," as ultimately it is responsible for the variation among individuals, populations, and species. The next level of the hierarchy is the *species* level: this is the level of the biodiversity hierarchy that most conservation legislation targets, where most conservation organizations focus their efforts, and what most people think of when they think of biodiversity.

The interactions between the individual organisms that make up a *population* (competition, cooperation, etc.), and their specializations for their environment (including ways in which they might modify the environment itself) are important aspects of the next levels of the biodiversity hierarchy. Interactions between different species (e.g., predator-prey relationships) and their environments form the next level of the hierarchy, focusing on *community* and *ecosystem* biodiversity. The largest scales of the biodiversity hierarchy are *landscapes* and *ecoregions*.

This hierarchy is one helpful way to organize the dizzying diversity of life, but it is important to keep in mind that there are other ways of portraying the various aspects of biodiversity. One framework focuses on the distribution of biodiversity over the earth's surface. The structure of communities and ecosystems (e.g., the number of individuals and

The Biodiversity Hierarchy

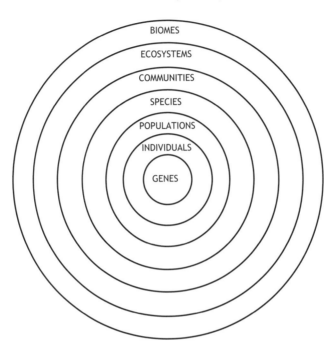

Figure 1.1 Biodiversity is sometimes represented as a hierarchy beginning with the diversity found in genes and extending to diversity at the continental level. Genes at the center of this hierarchy are the fundamental diversity upon which the variation at all other levels is based (*Cullman*)

species present) varies in different parts of the world. Similarly, the interactions between organisms in ecosystems and communities vary from one place to another. Different assemblages of ecosystems make up diverse landscapes. These spatial patterns of biodiversity are affected by climate, geology, and geography. For example, the numbers and types of species found in an arctic tundra ecosystem are significantly different from those found in a tropical wetland ecosystem. These concepts are covered in more detail in Chapter 2.

Another important aspect of biodiversity not explicit in the biodiversity hierarchy is its dynamism. Life is always changing and evolving; even ecosystems that seem static at first glance may be the products of long term dynamic processes. Certain changes occur periodically; for example, daily, seasonal, or annual changes in the species and number of organisms present in an ecosystem and how they interact. A deer may graze in one area and then bed down for the night in another. Atlantic salmon are spawned in freshwater streams, migrate out to sea to mature, and then return to freshwater to reproduce. Some ecosystems change in size or structure over time. A good example of this is the changes that forest ecosystems undergo with the presence of natural fires. Longleaf pine and scrub oak ecosystems of the southeastern United States are a good example of a fire-adapted ecosystem. With the periodic disturbance of a low-intensity fire, this forest type has an open canopy and a clear, park-like understory. Remove the fire, and the understory of the forest becomes dense with undergrowth and leaf litter. Certain ecosystems change seasonally. For example, vernal pools—depressions in the ground that flood seasonally—are important ephemeral habitat for frogs, nesting birds, invertebrates, and plants. Biodiversity also changes over longer, evolutionary time scales. Geological processes, such as the movement of the Earth's plates, mountain building and erosion, changes in sea level, and climate change cause significant, long-term changes to the structural and spatial characteristics of global biodiversity. For example, changes in sea level during different times sometimes exposed and sometimes obscured the Bering land bridge between what is now Russia and Alaska, facilitating and then impeding exchange of species between the two continents. Evidence of past sea level fluctuations—connecting and separating continents—is reflected in the evolutionary relationships between present-day species in Asia and North America. For example, the closest living relative of the towering Redwood trees of California (*Sequoia sempervirens*) are the Dawn Redwoods (*Metasequoia glyptostroboides*) found in present-day China, and the ancestors of the three extant species of Vietnamese glass lizards (a legless lizard that resembles a snake) are from North America.

ARE HUMANS PART OF BIODIVERSITY?

In Western culture, the dominant way of thinking considers humans to be separate and distinct from nature. Things are often described as being natural or man-made (created by the human mind), though the actual line between what is natural and what is artificial is often difficult to draw. This way of thinking can be traced to Western religious, literary, and historical traditions. For example, in the Bible, Adam was given dominion over the animals of the Garden of Eden. Writing by European colonists in North

America includes references to dominating the landscape, imposing control upon the chaos and disorder of the wilderness. Even Romanticism, the artistic and literary movement of the late eighteenth and early nineteenth centuries, which celebrated wilderness, did so from a vantage point: the emphasis was on the way that a view of nature engendered transcendental or spiritual experiences.

In our day-to-day life in the developed world, it is very easy to be ignorant of our connections to and our fundamental dependence on the natural world. Due to this cultural paradigm, many researchers confine biodiversity to natural variety and variability, excluding biotic patterns and ecosystems that result from human activity. In reality, however, it is difficult to assess the "naturalness" of an ecosystem because in most places human influence is pervasive, long-term, and varied. For example, while the Arctic may seem devoid of human activity relative to many places on Earth, many human-produced chemicals, such as mercury, lead, and PCBs, have reached Arctic environments. Recent research has also suggested that species and communities found in seemingly "untouched" ecosystems, such as the rainforests of the Amazon and Congo Basins, have been shaped by human communities in the past.

So, how to address this paradox? Some researchers consider humans to be a part of nature, and so they consider the cultural diversity of human populations and the ways that these populations use or interact with habitats and other species on Earth as a component of biodiversity too. Other researchers make a compromise between completely including or excluding humans as a part of biodiversity. They do not accept all aspects of human activity and culture as part of biodiversity, but they do recognize that the ecological and evolutionary diversity of domestic species, the species composition and ecology of agricultural ecosystems, and the landscape-level interactions between certain cultures and ecosystems are all part of biodiversity.

―――――――――― ∽०∾ ――――――――――

Genetic Diversity

The diversity of life we see on the planet is a product of genetic diversity, that is, the variation in the DNA (deoxyribonucleic acid) that makes up the genes of organisms leads to the variation we see at all other levels.

Genetic diversity refers to any variation in the nucleotides, genes, chromosomes, or whole genomes of organisms (the *genome* is the entire complement of DNA within the cells or organelles of the organism). Genetic diversity, at its most elementary level, is represented by differences in the sequences of four nucleotides (adenine, cytosine, guanine, and thymine), which form the DNA within the chromosomes in the cells of organisms. Some cells have specific organelles that contain chromosomes (for example, mitochondria and chloroplasts have their own chromosomes, which are separate from a cell's nuclear chromosomes). Nucleotide variation is measured for discrete sections of chromosomes, called "genes." Each gene comprises a hereditary section of DNA that occupies a specific place on the chromosome, and controls a particular characteristic of an organism. DNA provides the instructions to create proteins and in turn all other parts of a cell. Genetic diversity exists:

- within a single individual,
- between different individuals of a single population,
- between different populations of a single species (population diversity), and
- between different species (species diversity).

It is difficult, in some cases, to establish the boundaries between these levels of diversity. For example, it may be difficult to interpret whether variation between groups of individuals represents diversity between different species, or represents diversity only between different populations of the same species. Nevertheless, in general terms, these levels of genetic diversity form a convenient hierarchy for describing the overall diversity of organisms on Earth.

Most organisms have two sets of chromosomes (or are diploid), and therefore carry two copies of each gene, known as alleles. Some organisms have one, three, or even four sets of chromosomes, and are known as haploid, triploid, or tetraploid, respectively. While most animals are diploid, many plants, especially important crop plants such as wheat and bananas, have more than two sets of chromosomes. A gene or allele encodes for the production of amino acids that then string together to form proteins. Differences in the nucleotide sequences of alleles result in the production of slightly different proteins. These proteins lead to the development of traits or the specific anatomical and physiological characteristics that make up a particular organism; these in turn are responsible for determining aspects of an organism's behavior.

Variation between the alleles for each gene can be introduced through mutation, sexual reproduction, or through gene flow (as organisms move between different areas). Mutations are structural changes in an organism's genes that can be passed on to the next generation. For species that reproduce sexually (when reproduction requires two individuals), the offspring inherit one allele from each parent, whose alleles may be slightly different. This difference in alleles is more likely if parents come from different populations and gene pools. Also, when the offspring's chromosomes are copied after fertilization, genes can be exchanged in a process called sexual recombination, adding additional variation. Harmless mutations and sexual recombination may allow the evolution of new characteristics.

Besides having distinct combinations of genes, species may also vary in the total number of chromosomes present, as well as the shape and composition of the chromosomes. Examination of these chromosomal features is another way of describing genetic diversity. Eukaryotes, which include animals, plants, protists, and fungi, have cells with complex internal structures such as a nucleus, chloroplasts, and mitochondria. Most eukaryotic cells have linear chromosomes, either with four arms extending off a central point and thus approximating an X shape, some with three arms, or Y-shaped, and some with two arms. Prokaryotes (organisms that have cells without a nucleus including bacteria and archaea), as well as mitochondria have circular chromosomes.

Different species vary in the number of genes or chromosomes within their DNA or genome (see Table 1.1). A greater total number of coding areas in DNA does not necessarily correspond with a greater observable complexity in the anatomy and physiology of the organism, however. For example, the human genome is very small in comparison with the complexity of its expression—it is about the same size as some less complicated species, such as some bacteria species. In humans, however, more proteins are encoded per gene than in other species. The largest genome size recorded to date is that of the Marbled Lungfish (*Protopterus aethiopicus*), which has almost thirty-eight times the amount of DNA in its genome than humans do. Not all DNA actively codes for proteins; some DNA act as a kind of placeholder, or as an end piece, to protect

Table 1.1 A Comparison of Genome Sizes. Genome Sizes Do Not Correspond Directly to Complexity

Species	Common Name	Number of Genes in Genome
Arabidopsis thaliana	thale cress—a plant	25,498
Oryza sativa	Indian rice	46,022–55,615
Caenorhabditis elegans	nematode	19,000
Drosophila melanogaster	fruit fly	13,600
Homo sapiens	human	*ca.* 30,000–40,000

the coding DNA during protein building. Some species have a lot more of this noncoding DNA than other species. Scientists still cannot explain all the reasons why some genomes have so much active DNA and others so little.

Analyses of genetic diversity can be applied to studies of the evolutionary ecology of populations. Genetic studies can identify alleles that might affect the ability of an organism to survive in its existing habitat, or might enable it to survive in more diverse habitats. Heritable characteristics form the basis for "natural selection"—some alleles or genes confer a selective advantage on an organism, making it more likely to survive than if it did not have them.

The presence of unique genetic characteristics distinguishes members of a given population from those of other populations. Large populations usually have a greater diversity of alleles than small ones. This diversity of alleles indicates a greater potential for the evolution of new combinations of genes and, subsequently, a greater capacity for evolutionary adaptation to different environmental conditions. In small populations, individuals are likely to be genetically, anatomically, and physiologically more alike than in larger ones and thus less adaptable to changing environmental conditions and more susceptible to disease. This situation of low genetic diversity is sometimes known as the "inbreeding effect." Maintaining genetic diversity is a key component of conservation efforts. Cheetahs (*Acionyz jubatus*) and the northern Elephant Seal (*Mirounga angustirostris*) are famous for having populations with low genetic diversity. But low genetic diversity is a concern for many endangered species from the Mediterranean monk seal (*Monachus monachus*) to the Northern hairy-nosed wombat (*Lasiorhinus krefftii*).

Phenotypic Diversity

Genes code for what something looks like or the outward expression of physical traits of an organism; however, the environment also modifies the way genes are physically expressed in the organism. The physical appearance of an organism results from its genetic makeup or *genotype,* and the action of the environment on the expression of the genes is termed its *phenotype.*

Phenotypic variation, therefore, refers to the variation of the physical traits, or phenotypic characters of the organism, such as differences in anatomical, physiological, biochemical, or behavioral characteristics. As noted above, phenotypic diversity is, in part, a product of genetic diversity and in part the influence of the environment on gene expression.

Nevertheless, the phenotypic characters represent an important measure of the adaptation of an organism to its environment, because it is these phenotypic characters that interact with the living and nonliving parts of the environment.

Phenotypic diversity between individuals, populations, and species is usually described in terms of the variation in the external morphology or the outward appearance of individuals. Variation in physiological and biochemical characteristics of the organism are also important indicators of phenotypic diversity. Behavioral characteristics represent the way in which an organism interacts with its environment, and are a product of the genes, which specify particular anatomical, physiological, or biochemical traits that might be adaptations for the environment.

Studies of blue-footed boobies (*Sula nebouxii*) on the Galapagos Islands, illustrate how environmental and genetic factors combine to influence behavior. Blue-footed boobies often lay broods of two, but one egg develops first and hatches before the other. Sometimes the first-born chick will push the other chick out of the nest when it is born, ensuring its death by malnutrition or heat exposure. Researchers from Wake Forest University in North Carolina have found that there is a genetic basis for the sibling-killing behavior, but that the parents control whether or not the first chick will kill its sibling. If there is enough food to rear both siblings, the parent boobies suppress the behavior. If food is scarce, however, the parents will not interfere. Thus, variation in behavioral characteristics may also be used to describe phenotypic diversity between individuals, populations, or species.

Local environmental conditions can alter phenotypic characters. In plants, leaf shape varies significantly among individuals of the same species occupying different habitats (e.g., dry versus wet sites, or sunny versus shaded sites). In a classic study of human twins separated at birth, the twin who was raised in a high altitude environment had a bigger heart and lungs than the one raised at sea level. Thus, any discussion of phenotypic diversity should account for the interrelationships between anatomical structure and function for organisms living in different habitats.

Indeed, phenotypic characters are the product of complex interrelationships between the form and function of various body tissues and organs. For example, because of the physical properties of light, vertebrate eyes must be a minimum size to be functional. Consequently, the eyes of tiny vertebrates (such as species of amphibians and fishes) are disproportionately large to their bodies. This, in turn, may affect the development and function of adjacent organs in the head, where there is a "competition" for headspace.

The extent to which genetic variation between organisms is expressed in their phenotypes can be quite variable for different characteristics. Genetic variation between some features might be expressed as very subtle differences in their phenotype. For example, populations and subspecies of the herring gull (*Larus argentatus*) and the lesser black-backed gull (*Larus fuscus*) are distinguished by very slight differences in the coloration of individuals. In some cases these differences are difficult to detect. However, genetic variation within a species can be quite extensive, particularly in cultured plants or domesticated animals where particular features have been artificially selected in different strains or breeds. For example, broccoli, cabbage, and cauliflower look very different from one another, but are all varieties of *Brassica oleracea* (see Figure 1.2). Another everyday example is breeds of dogs: Chihuahuas and Doberman pinschers, though having very different outward appearances, are both the same species, *Canis lupus familiaris*.

Population Diversity

A population is a group of the same species with a shared characteristic, usually living in the same area. Scientists also differentiate populations by their breeding rates and migration patterns, among other characteristics. The species, *Puma concolor*, for example, is known by many common names, including mountain lion, cougar, puma, catamount, because the large cat lives in wild areas from the southwestern United States to the Andes of South America. About thirty-two subspecies of *Puma concolor* have been described based on geographic location and differences in size and form; for example, their size varies from about 50 to 70 kg (100 to 155 lbs.) at the equator and reaching twice that size at the northern extent of their ranges (Canada's Yukon and Argentina's Patagonian pampas). Recent genetic analyses divide puma subspecies into six main groups, largely separated by geographic barriers that have restricted gene flow (or the transfer of genes between populations via migration).

Population diversity may be measured by the variation in genetic and morphological features that define different populations. This diversity may also be measured in terms of a population's demographics, such as the number of individuals present or the proportion of different ages or sexes in the population. However, it is difficult to measure demography and genetic variation (e.g., allele frequencies). A more practical way of

Figure 1.2 Even in one species there can be an incredible morphological diversity. Brussels sprouts (upper left), cabbage (center), kohlrabi (upper right), cauliflower (lower right), broccoli (lower left) are all the same species Brassica olearacea. They represent the following varieties: gemmifera, capitata, gonglyodes, botryis, and botryis (*Frey © CBC-AMNH.* Used by permission)

defining a population and measuring its diversity is to simply determine the area that a population occupies. Using this criterion, a population is a group of individuals of the same species occupying a particular area at the same time. The area occupied is usually defined based on ecological needs important to the species: for example, a particular pond or lake for a population of fish, or a south-facing slope for a population of light-loving trees.

The geographic range and distribution of populations (or their spatial structure) represent key factors in analyzing population diversity because they give an indication of the likelihood of movement of individuals between populations, and consequently of genetic and demographic interchange. Similarly, an estimate of the overall population size provides a measure of the potential genetic diversity within the population; large populations usually represent larger gene pools and hence greater potential diversity (see "Genetic Diversity"). Even though many species of sea turtles are wide ranging, their populations are segmented, depending in part on their tolerance for different temperatures and habitat preference. Among the world's seven species of sea turtles, the cold-tolerant leatherback's (*Dermochelys coriacea*) populations in the Indo-Pacific and Atlantic are genetically connected, while the tropical green (*Chelonia mydas*), hawksbill (*Eretmochelys imbricata*), and olive and kemp's ridleys (*Lepidochelys olivacea vs. L. kempi*) have ancient separations between oceans.

Isolated populations, with very low levels of interchange, may show high levels of genetic divergence, and exhibit unique adaptations to the biotic and abiotic characteristics of their habitat. The genetic diversity of groups that generally do not disperse well—such as amphibians and some herbaceous plants—may be restricted to local populations. For this reason, range retractions of species leads to loss of local populations and the genetic diversity they hold. Loss of isolated populations, along with their unique genetic variation, is considered by some scientists to be one of the greatest but most overlooked tragedies of the biodiversity crisis. Isolated populations of species are basically incubators for new species; when they disappear, the potential for the evolution of a new species also disappears.

Populations can be categorized according to the level of divergence between them. Isolated and genetically distinct populations of a single species are often called subspecies. Animal populations that show less genetic divergence might be recognized as "variants" or "races." For plants, the ranks below species include subspecies, variety, subvariety, form, and subform. However, the distinctions between subspecies and other categories can be somewhat arbitrary (see below for further discussion under "Species Diversity").

A species with specialized habitat needs may assume the same distribution as that particular habitat type. Populations of that species are likely to be confined within patches of that habitat type. For example, alpine species may be confined to particular mountaintops, and each mountaintop would represent a population of that species. This is the case, for example, for species that live in small wetlands, springs, caves, particular soil or forest types, and many other comparable situations. Periodically, individual organisms may disperse from one population to another, facilitating genetic exchange between the populations. This group of different but interlinked populations, with each population located in its own, discrete patch of habitat, is called a *meta-population*.

Species Diversity

This is the level of biodiversity that most of us are aware of on a day-to-day basis, whether through information from environmental organizations, the news, or the Discovery Channel. While what exactly defines a species is a matter of debate among scientists (see description of different species concepts below), much environmental management and political attention focuses on individual species. The Endangered Species Act of the United States, which aims to protect species richness, is one example.

Species diversity has two components: the number of different species in a particular area, and the relative abundance of individuals within different species in the same area. This first component is known as *species richness*, while the second is known as *species evenness*. An ecosystem in which all the species are represented by the same number of individuals has high species evenness. In contrast, an ecosystem in which some species are represented by many individuals, and other species are represented by very few individuals, has low species evenness. *Species diversity* encompasses both species richness and evenness. It is common, though incorrect, for people to use the term "species diversity" interchangeably with species richness.

Species diversity is often described in terms of the *phylogenetic diversity*, or evolutionary relatedness, of the species present in an area. (Note that phylogenetic diversity also exists at the genus and family level.) For example, some areas may be rich in closely related taxa, having evolved from a common ancestor that was also found in that same area, whereas other areas may have an array of less closely related species descended from different ancestors.

Scientist and conservation managers use diversity indices to quantitatively describe species diversity. One of the most well known is the Shannon index, which incorporates species richness (or the number of species) and species evenness (the abundance of each species) to measure the diversity of an area. In a typical acre (4,047 sq. m.) of most northern forests, there are about a dozen tree species. For example, a mixed-hardwood forest in the northeastern United States is dominated by beech and maple trees, with birch and ash trees at lower densities. These forests have high species evenness, but low overall species richness. In contrast, tropical rainforests have hundreds of species per acre, but each one at a low abundance. Thus, they have high species richness but low species evenness.

What Is a Species?

To count the number of species, we must define what we mean by a *species*. There are several competing ways of defining a species. While this might seem like a strange topic to disagree on, we hope that the following explanation will reveal some of the reasons scientists have not reached a consensus on how to define a species. After all, natural selection and evolutionary theory predicts that species are not static entities. Species evolve and change, sometimes into new and different species. How does one decide when two populations are different enough to constitute two species? The different species concepts (morphological, biological, and phylogenetic) outlined below answer this question in different ways. The concepts center on ideas of genetic relatedness, or ancestry, among species. These concepts are not congruent, and considerable debate exists about their advantages and disadvantages. Different models are typically useful for different applications. Just as the behavior of light cannot be explained entirely by either the wave or the photon model, individual species concepts cannot explain all of the ways of thinking of how to distinguish between species.

The *morphological species concept* is the oldest of the approaches currently used to answer the question, what is a species, and also the most readily understandable. According to this concept, individuals that look alike and share the same identifying traits belong to the same species. While this is an outdated concept, this is still the method that biologists initially use to distinguish one species from another in a field setting: maple trees have leaves with triangular lobes; Thompson's gazelles have a white patch bisected by a black stripe on their sides; and Queen Anne's lace has small, white, composite flowers. One flaw of this concept is that it is really a matter of opinion as to how similar two individuals must be in order to count them as the same species. This method is a useful way to generally categorize species, but its application is limited at finer levels. Genetic analyses are leading to a reevaluation of the concept as discoveries reveal that individual animals that superficially look the same are actually different species. According to the scientific method, scientific conclusions should be based on reproducible and verifiable data, which can be arrived at no matter who follows the method. The morphological species concept relies too heavily on accumulated experience rather than deductive methods for distinguishing between species and thus it is too subjective, so scientists have searched for other ways to describe species.

The *biological species concept*, defines species as a group that interbreeds and is isolated from other groups. Basically, two individuals are the same species if they can breed and produce viable offspring—that is offspring that can also breed. This approach is an improvement on the morphological species concept because it uses a testable characteristic to differentiate between species. For example, when a female tiger and a male lion mate they produce a "liger." Ligers are not able to breed with each other, and produce other ligers, therefore it is clear that tigers and lions are different species. Similarly when a male donkey and a female horse mate they produce a mule that is also unable to reproduce. Problems with this approach to defining a species is that it is difficult to apply to organisms that do not reproduce sexually or to organisms that do not normally live in the same place or time, so it is impossible to know whether they would interbreed.

According to the *phylogenetic species concept*, a species is a group of individual organisms that share a common ancestor. This approach is more complex to apply than the morphological approach, and recognizes more species than the biological species concept. However, it is particularly helpful for testing scientific hypotheses.

In practice, systematists (scientists that study how different species are evolutionarily related) use a combination of species concepts. They usually group specimens together according to shared features or characters; these could be genetic, morphological, or physiological characters. When two or more groups show different sets of shared characters, and the shared characters for each group allow all the members of that group to be distinguished relatively easily and consistently from the members of another group, then the groups are considered different species. This approach relies on the objectivity of the phylogenetic species concept (i.e., the use of intrinsic, shared, characters to define a species) and applies it to the practicality of the morphological species concept, in terms of sorting specimens into groups.

An *"evolutionary significant unit"* (ESU) is defined as a group of organisms that has undergone significant genetic divergence from other groups of the same species. Identifying ESUs requires natural history information, range and distribution data, and a suite of genetic analyses. If the ESUs are based on populations that are *sympatric* (i.e., occupying the same geographic area), or *parapatric* (i.e., occupying adjoining but not overlapping ranges), then it is particularly important to give evidence of significant genetic distance between those populations.

ESUs are important for conservation management because they can be used to identify discrete components of the evolutionary legacy of a species that warrant conservation action. Nevertheless, in evolutionary terms and hence in many systematic studies, species are recognized as the minimum identifiable unit of biodiversity above the level of a single organism. Thus, there is generally more systematic information available for species diversity than for subspecific categories and for ESUs. Consequently, estimates of species diversity are used more frequently as the standard measure of overall biodiversity of a region.

Despite their differences, all species concepts are based on the understanding that there are parameters that make a species a discrete and identifiable evolutionary entity. If populations of a species become isolated, either through differences in their distribution (i.e., geographic isolation) or through differences in their reproductive biology (i.e., reproductive isolation), they can diverge, ultimately resulting in the creation of new species. During this process, we expect to see distinct populations representing "incipient species"—species in the process of formation. Some researchers describe these as subspecies, or some other subcategory, according to the species concept used. However, it is very difficult to decide when a population is sufficiently different from other populations for it to be ranked a subspecies. For these reasons, deciding if a population merits subspecific and infra-subspecific ranks may become extremely subjective decisions. This becomes particularly important for conservation decisions. Are we concerned when a small population, such as the Florida panther, disappears, if we know there are other panther populations still present elsewhere? If we consider a Florida panther to be distinct enough then it does matter. Depending on which species concept is used, there are different conservation priorities. Some species concepts will recognize Florida panthers as a distinct subspecies and thus a high conservation priority, other species concepts will consider them too similar to panthers elsewhere and thus Florida panthers are a lower conservation priority.

How Many Species Are There?

Estimates of the total number of species in the world are based on extrapolations from what we already know about certain groups of species. For example, we can estimate the total number of species in a particular group of organisms by extrapolating from the ratio of scientifically described species to undescribed species of that group that were collected from a particular area. While we know a lot about birds and mammals, we know very little about other groups, such as bacteria and fungi, for which we do not have suitable baseline data from which we can extrapolate our estimated total number of species on Earth. Additionally, some groups of organisms have not been comprehensively collected from areas where their species richness is likely to be richest (for example, insects in tropical rainforests). These factors, and the fact that different people have used different techniques and data sets to extrapolate the total number of species, explain the large range between the lower and upper estimates of 3.6 million and 117.7 million, of the total number of species.

More significantly, some species are very difficult to identify. For example, taxonomically "cryptic species" look very similar to other species and may be misidentified (and hence overlooked as being a different species). Thus, several different but similar-looking species, identified as a single species by one scientist, are later identified as a completely different species by another scientist. As taxonomists develop more techniques to uncover genetic and phenotypic diversity, they provide more reliable estimates of the number of species on the planet.

CRYPTIC CASCADE FROGS

Cascade frogs of the *Rana chloronota* complex live in fast-moving montane streams and nearby forested areas in various parts of Southern Asia (see Figure 1.3). Until recently the three species of cascade frogs that make up the *Rana chloronata* complex were thought to be one species as they were very similar looking. These are often termed *cryptic species.*

Rana chloronota is a widespread species that is found in montane waterways of Central and Northern Vietnam, and at the same latitudes in China, Myanmar, and India. *Rana morafkai* was recently described from Vietnam's Central Highlands, but has also been documented in Lao PDR. *Rana banaorum* has only been found in the Central Highlands of Vietnam and only described in 2003. Since these species are very similar looking, their exact ranges and the relationships among the species are still unclear.

What Kinds of Species?

When asked what he had learned about God after studying nature, British scientist Lord Haldane famously replied that he must have had "an inordinate fondness for beetles" (Hutchinson 1959, p. 146)

Figure 1.3 Cryptic species look so similar to other species that they are easily misidentified. These cascade frogs of the Rana chloronota complex are one example. Once considered to be a single species, they represent three different species: *Rana chloronota, Rana morafkai,* and *Rana banaorum* (*Bain©CBC-AMNH.* Used by permission)

Our understanding of species reflects human biases toward species that are more like us or are important to us. So, most people's (and scientist's) attention has focused on large, charismatic, and economically important species, such as mammals, birds, and certain trees (e.g., mahogany, sequoia) and fish (e.g., salmon). However, the majority of Earth's species are found in other, generally overlooked groups, such as bacteria, mollusks, insects, and flowering plants. The largest group of described species is the Arthropods, with about 1.5 million identified to date; they make up about 75 percent of known species. Arthropods include insects, spiders, crustaceans (e.g., crabs and lobsters), and centipedes. Though often forgotten, these groups play important functional roles in ecosystems, and future scientific effort will continue to uncover their roles.

The proportional representation of different groups of related species (e.g., bacteria, flowering plants, insects, birds, mammals) is usually referred to as taxonomic or phylogenetic diversity. Species are grouped together according to shared characteristics (genetic, anatomical, biochemical, physiological, or behavioral), which leads to a classification of species based on their apparent evolutionary or ancestral relationships. We then use this information to assess the proportion of related species with respect to the total number of species on Earth (see Tables 1.2 and 1.3).

Table 1.2 Estimated Number of Species by Taxonomic Group

Taxon	Taxon Common Name	Number of Species Described $(N)^*$	N as Percentage of Total Number of Described Species*
Bacteria	true bacteria	9,021	0.5
Archaea	archaebacteria	259	0.01
Stramenopiles	stramenopiles	105,922	6.1
Angiospermae	flowering plants	233,885	13.4
Other Plantae	(eg., red algae, mosses, ferns, conifers)	49,530	2.8
Fungi	fungi	100,800	5.8
Mollusca	mollusks	117,495	6.7
Nematoda	nematodes	20,000	1.1
Arachnida	arachnids	74,445	4.3
Crustacea	crustaceans	38,839	2.2
Insecta	insects	827,875	47.4
Other Invertebrate Metazoa	(e.g., cnidarians, platyhelmiths, annelids, echinoderms)	82,047	4.7
Actinopterygii	ray-finned bony fishes	23,712	1.4
Other Vertebrata	(e.g., amphibians, reptiles, mammals, birds)	27,199	1.6
Other Eucarya	(e.g., alveolates, eugelozoans, choanoflagellates)	36,702	2.1

* The table includes representatives from the three recognized domains of Archaea, Bacteria, and Eucarya. Examples of eucaryan groups are restricted to those that include at least 1 percent of the total number of described species on Earth. The total number of species is estimated at 1,747,851.

Source: This figure, and the numbers of species for representative taxa are based on Lecointre and Guyader (2001).

Table 1.3 Of the 1.7 million species described, the vast majority of them are insects

Taxonomic Group	Proportion of Total Described Species
Insects	47.3%
Flowering Plants	13.4
Mollusks	6.7
Stramenopiles	6.1
Fungi	5.8
Other Invertebrates (cnidarians, annelids, echinoderms, platyhelmiths)	4.7
Arachnids	4.3
Other Plants (red algae, moss, fern, conifers)	2.8
Crustaceans	2.2
Other Eucaryotes (choanoflagellates, eugelozoans, alveolates)	2.1
Other Vertebrates (includes all species of birds, mammals, reptiles and amphibians)	1.6
Ray-finned bony fishes	1.4
Nematodes	1.1
Bacteria	0.5
Archaebacteria	0.1

----------------------------------- ﹏ﻪﻤﻤ -----------------------------------

MEASURING BIODIVERSITY

Measuring biodiversity is not a straightforward task. Not only is it difficult to define what a species is, but it is often also a challenge to sample and identify species. Before 1982, most scientists estimated that the number of species on the planet was in the order of millions. It was a new approach to sampling beetles that led to a reevaluation of how many species exist. Forest canopies (vegetation and branches of trees above ground, including the crowns of trees and any plants growing on them) were believed to host an incredible diversity of beetles but are difficult to access. In fact, most forest species live in the canopy. An innovative sampling technique of tree canopies pioneered by scientist Dr. Terry Erwin in 1982 led to a major revision of the number of beetle species on the planet. Erwin fogged the tree canopy with a pesticide, catching all the insects below. Erwin sampled nineteen trees of one species (*Luehea seemannii*) over the course of three seasons and found over 1,200 different beetle species. He used these discoveries to calculate the total number of arthropods on the planet. Erwin calculated the number of specialized beetles per tropical tree species as 165. As beetles make up about 40 percent of arthropods, he then came up with an estimate of specialized arthropods and multiplied this by the number of tropical tree species (~50,000) to get to an estimate of 30 million tropical insects. (Essentially Erwin used tree species richness as a proxy for insect diversity.). This led to a revision of the number of species on the planet from millions to tens of millions, though some estimates are considerably higher. However, most biologists think some of Erwin's assumptions were incorrect, for example the number of specialized insects has been reduced by a factor of four, and led to an overestimate of the number of undescribed species. Erwin's study nevertheless led to a major increase in the number of species believed to exist on the planet compared to estimates before his study. Even today no one really knows the exact number of species on the planet and some of the smallest species (bacteria) are only now being studied more intensely.

――――――――――――――――― ∽⃝ ―――――――――――――――――

Community Diversity

A *community* comprises *populations of different species* that naturally occur and interact in a particular environment. Some communities are relatively small in scale and may have well-defined boundaries. Some examples are: species found in or around a desert spring, the collection of species associated with ripening figs in a tropical forest, those clustered around a hydrothermal vent on the ocean floor, those in the spray zone of a waterfall, or under warm stones in the alpine zone on a mountaintop. Other communities are larger, more complex, and may be less clearly defined, such as old-growth forests of the northwest coast of North America, or lowland fen communities of the British Isles.

Like the term "population," "community" has some flexibility in the way it is used by biologists. Sometimes biologists apply the term "community" to a subset of organisms within a larger community. For example, some biologists may refer to the "community" of species specialized for living and feeding entirely in the forest canopy, whereas other biologists may refer to this as part of a larger forest community. This larger forest community includes those species living in the canopy, those on the forest floor, and those moving between these two habitats, as well as the functional interrelationships between all of these. Similarly, some biologists working on ecosystem management might distinguish between the community of species that are endemic to an area (e.g., species that are endemic to an island) as well as those "exotic" species that have been introduced to that area. The introduced species form part of the larger, modified community of the area, but might not be considered as part of the region's original and distinctive community. Each scientist's definition and delineation of "community" reflects his or her own focus and does not necessarily invalidate the others.

Communities are frequently classified by their overall appearance, or physiognomy. For example, coral reef communities are classified according to the appearance of the reefs where they are located, that is, fringing reef communities (run parallel and less than 1 km off the shore), barrier reef communities (also run parallel but are 5 km offshore), and atoll communities (a ring of coral circling a shallow lagoon). Similarly, different stream communities may be classified by the physical characteristics of the stream where the community is located, such as riffle zone and pool communities. However, one of the easiest, and hence most frequent methods of community classification is based on the dominant species present—for example, intertidal mussel bed communities, Ponderosa pine forest communities of the Pacific northwest region of the United States, or Mediterranean scrubland communities.

The factors that determine the diversity of a community are extremely complex. There are many theories on what these factors are and how they influence community and ecosystem diversity. Environmental factors, such as temperature, precipitation, sunlight, and the availability of nutrients, are very important in shaping communities and ecosystems. One way of measuring community diversity is to examine the energy flow through food webs that unite the species within the community; the number of links in the food web provides a quantitative measure of the extent of community diversity. However, in practice, it is very difficult to quantify the energy exchange that occurs due to all of the many interactions among species within a community, particularly if one includes microbes in the analysis. It is easier to measure the genetic diversity of the populations in the community, or to count the numbers of species present, and use these as proxies to describe community diversity. The evolutionary diversity of the species present is another proxy for measuring the community diversity.

Ecosystem Diversity

An *ecosystem* is a *community plus the physical environment* that it occupies at a given time. An ecosystem can exist at any scale, for example, from the size of a small tide pool up to the size of the entire biosphere. However, lakes, marshes, and coral reefs, forest stands represent more typical examples of the areas that are commonly compared in discussions of ecosystem diversity. Ecosystems may be classified according to the dominant type of environment, or the dominant type of species present; for example, a salt marsh ecosystem, a rocky shore intertidal ecosystem, a mangrove swamp ecosystem. Because temperature is an important aspect in shaping ecosystem diversity, it is also used in ecosystem classification (e.g., cold winter deserts, versus warm deserts).

Broadly speaking, the diversity of an ecosystem depends on the physical characteristics of the environment, the diversity of species present, and the interactions that the species have with each other and with the environment. Therefore, the functional complexity of an ecosystem can be expected to increase with the number and taxonomic diversity of the species present, and the vertical and horizontal complexity of the physical environment. However, one should note that some ecosystems (such as hydrothermal vents, or hot springs), which do not appear to be physically complex or especially rich in species, might be considered to be *functionally complex*. This is because they include species that have remarkable biochemical and behavioral specializations for surviving in harsh environments and obtaining their energy from inorganic chemical sources.

The physical characteristics of an environment that affect ecosystem diversity are themselves quite complex. These characteristics include, for example, the temperature, precipitation, and topography of the ecosystem. There is a general trend for warm and moist tropical ecosystems to be richer in species than cold temperate ecosystems (see section on "Spatial Gradients in Biodiversity"). Also, the energy flux in the environment significantly affects the ecosystem. An exposed coastline with high wave energy will have a considerably different type of ecosystem than a low-energy environment such as a sheltered salt marsh. Similarly, an exposed hilltop or mountainside is likely to have stunted vegetation and low species diversity compared to more prolific vegetation and high species diversity found in sheltered valleys.

While the physical characteristics of an area will significantly influence the diversity of the species within a community, the organisms in turn also modify the physical characteristics of the ecosystem. For example, stony corals (Scleractinia) build extensive calcareous skeletons that are the basis for coral reef ecosystems that extend thousands of kilometers, as is the case for the Great Barrier Reef in Australia. Organisms also modify their ecosystems in more subtle ways. Trees modify the microclimate as well as the structure and chemical composition of the soil around them. When pine needles decompose, for example, the soil becomes more acidic, which limits the plant species that can establish there.

Environmental disturbance on a variety of temporal and spatial scales affects the species richness and, consequently, the diversity of an ecosystem. For example, river systems in the North Island of New Zealand have been affected by volcanoes several times over the last 25,000 years. Eruptions led to ash-laden floods, which killed most of the fish in the rivers, and recolonization has been possible by a limited number of species. Once the physical environment of the disturbed rivers recovered, diadromous fish (or those that migrate between freshwater and seawater at fixed times during their life cycle, such as salmon and eels) were able to recolonize the rivers by traveling from other unaffected rivers via the sea, but the species richness of the disturbed rivers remains low. Nevertheless, occasional disturbance can have the reverse effect, and increase the species richness of an ecosystem by creating spatial heterogeneity in the ecosystem, allowing new species to colonize or by preventing certain species from dominating the ecosystem.

Landscape Diversity

A *landscape* is made up of a collection of common land forms, vegetation types, and land uses. Therefore, *assemblages of different ecosystems* (the physical environments and the species that inhabit them, including humans) create *landscapes* on Earth. Although there is no standard definition of the size of a landscape, they are usually on the order of hundreds or thousands of square kilometers (tens or hundreds of square miles, or tens to hundreds of thousand acres). The landscape level of biodiversity is a relatively new horizon for scientific research due to technological innovations in analyzing satellite images and geographic information systems (GIS) software. The study of landscapes is often closely tied to land use planning and human use of land.

Species composition and population viability are often affected by the structure of the landscape; for example, the size, shape, and connectivity of individual patches of ecosystems within the landscape. Conservation management ideally should be directed at whole landscapes to ensure the survival of species that range widely across different ecosystems (e.g., jaguars, quetzals, or species of plants that have widely dispersed pollen and seeds).

Diversity within and between landscapes depends on local and regional variations in environmental conditions, as well as the species supported by those environments. Landscape diversity is often incorporated into descriptions of "ecoregions," as discussed below.

Ecoregions

Since the 1980s, there has been an increasing tendency to map biodiversity over ecological regions or "ecoregions." Ecoregions include ecosystems that share certain distinct characters. An *ecoregion* is a large area of land or water (typically spanning millions of acres or thousands of square kilometers) with a geographically distinct collection of species and natural communities. Several standard methods of classifying ecoregions have been developed, with landform, climate, altitude, ecological processes, and predominant vegetation being important criteria.

The USDA Forest Service classification, developed by Robert Bailey for the United States, is one of the most widely adopted methods for designating ecoregions. It is a hierarchical system with four levels: domains, divisions, provinces, and sections. *Domains* are the largest geographic levels and are defined by climate: polar, dry, humid temperate, and humid tropical. Domains are split into smaller *divisions* defined according to climate and vegetation, and the divisions are split into smaller *provinces* that are usually defined by their major plant formations (see Table 1.4 for some examples). Some divisions also include

Table 1.4 Biomes and Ecoregions of the United States

Domain	Division	Province Mountain Province	Section
Polar (100)	Tundra (120)	Bering Tundra (Northern) (125)	Kotzebue Sound Lowlands (125A)
		Seward Peninsula Tundra—Meadow (M125)	Seward Mountains (M125A)
	Subarctic (130)	Upper Yukon Tayga (139)	Upper Yukon Flats (139A)
		Upper Yukon Tayga—Meadow (M139)	Upper Yukon Highlands (M139A)
Dry (300)	Tropical/ Subtropical Desert (320)	Chihuahuan Semidesert (321)	Basin and Range (321A)
			Stockton Plateau (321B)
		American Semidesert and Desert (322)	Mojave Desert (322A)
			Sonoran Desert (322B)
			Colorado Desert (322C)
	Temperate Steppe (330)	Great Plains Steppe (332)	Northeast Glaciated Plains (332A)
			Nebraska Sand Hills (332C)
		Middle Rocky Mountain Steppe—Coniferous Forest—Alpine Meadow (M332)	Idaho Batholith (M332A)
			Rocky Mountain Front (M332C)

Source: Based on Bailey 1983, 1998; McNab and Avers 1994.

varieties of "mountain provinces." These generally have a similar climatic regime to the neighboring lowlands but show some altitudinal zonation, and they are defined according to the types of zonation present. Provinces are divided into *sections*, which are defined by the landforms present.

The ecoregion approach is a practical unit for conservation planning; the hierarchical nature of Bailey's ecoregion classification allows for conservation management to be implemented at a variety of geographical levels, from small-scale programs focused on discrete sections, to much larger national or international projects that target divisions. The ecoregion approach is used by both the World Wildlife Fund (WWF) and The Nature Conservancy (TNC) for conservation planning, as detailed in Chapter 2.

Biodiversity over Time

The evolutionary history of Earth has physically and biologically shaped our contemporary environment. As noted in the section on *Biogeography*, plate tectonics and the evolution of continents and ocean basins have been instrumental in directing the evolution and distribution of the Earth's biota. The physical environment has also been extensively modified by these biota. Many existing landscapes are based on the remains of earlier life forms. For example, some existing large rock formations are the remains of ancient reefs formed 360 to 440 million years ago by communities of algae and invertebrates. Very old communities of subterranean bacteria may have been responsible for shaping many geological processes during the history of the Earth, such as the conversion of minerals from one form to another, and the erosion of rocks. The evolution of photosynthetic bacteria, sometime between 3.5 and 2.5 million years ago, played an important role in the evolution of the Earth's atmosphere. These bacteria released oxygen into the atmosphere, changing its composition from 1 percent oxygen to the one we know today, where oxygen makes up 21 percent of the atmosphere. It probably took over 2 billion years for the oxygen concentration to reach the level it is today. The process of oxygenation of the atmosphere led to important evolutionary changes. Organisms developed mutations that enabled them to use oxygen as a source of energy. The rise of animal and plant life on land was associated with the development of an oxygen-rich atmosphere.

Extinction

Extinction, or the complete disappearance of a species from Earth, is an important part of the evolution of life. The current diversity of species is a product of the processes of extinction and speciation throughout the 3.5 billion year history of life. Assuming there are about 40 million species alive today, and between 5 and 50 billion species have lived at some time during the history of the Earth, then 99.9 percent of all the life that has existed on Earth is now *extinct* (Raup 1991). Most (if not all) species eventually become extinct. The average duration of a species in the fossil record [or the species turnover time] is a few million years. Most vertebrates last about 1 to 3 million years, while invertebrates last about 10 to 30 million years. Of course, there are some species that have defied the odds; these ancient species are often called "living fossils" for their ability to survive unchanged over long periods of time (see Table 1.5). The horseshoe crab is even more ancient than once thought. Recent fossils discovered in Manitoba in 2007 indicate it originated 450 million years ago. It is the oldest marine arthropod, actually more closely related to spiders than crabs. It is still widespread; besides the species in the Atlantic Ocean, three species in the same family (*Limulidae*) are found in the Indian and Pacific Oceans, it is particularly common in the Gulf of Mexico.

Table 1.5 These "Living Fossils" Have Survived Unchanged against the Odds. Most Vertebrates Survive 1 to 3 Million Years on Earth, While Invertebrates Last About 10 to 30 Million Years

Common Name/Scientific Name	Origins	Current Distribution
Horseshoe crab (*Limulus polyphemus*)	450 million years	Atlantic Ocean; three species in the same family (*Limulidae*) are found in the Indian and Pacific oceans
Gingko or Maidenhair Tree (tree) (*Gingko biloba*)	170 million years	Eastern China
Dawn Redwood (tree) (*Metasequoia glyptostroboides*)	100 million years	remote valley in Central China
Wollemi Pine (*Wollemia nobilis*)	110 million years	Australia
Gladiator Insects (Mantophasmatodea—12 species in the Order)	45 million years	Southern Africa
Coelacanth (fish) (*Latimeria spp.*)	410 million years	Madagascar, East Africa, Indonesia
Tuatara (related to lizards and snakes) (*Spenodon spp.*)	220 million years	New Zealand
Chambered Nautilus (*Nautilus pompilius*)	500 million years	Andaman Sea and southern Pacific, west to the island of Fiji

Source: Compiled by Laverty.

Table 1.6 Geological Timeline

Duration (millions of years)	Era	Period		Events
3,800 to 570	Precambrian			Bacteria and blue-green algae
570 to 500	Paleozoic	Cambrian		First invertebrates
500 to 440		Ordovician		
440 to 410		Silurian		First land plants First land invertebrates
410 to 365		Devonian		First ferns and seed plants First insects First amphibians
365 to 290		Carboniferous		First flying insects First reptiles
290 to 245		Permian		
245 to 210	Mesozoic	Triassic		Extinction of trilobites First dinosaurs First mammals
210 to 140		Jurassic		First birds
140 to 65		Cretaceous		First flowering plants First primates Extinction of dinosaurs Extinction of ammonites
65 to 55	Cenozoic	Tertiary	Paleocene	Diversification and spreading of mammals
55 to 38			Eocene	
38 to 25			Oligocene	First grass
25 to 5			Miocene	
5 to 2			Pliocene	First hominids
0.01–2		Quarternary	Pleistocene	
Present–0.01			Holocene	

Source: Compiled by Laverty.

Extinction has not occurred at a constant pace through the Earth's history. There have been at least five *mass extinctions,* periods when the extinction has more than doubled, and the taxa affected have included representatives from many different taxonomic groups of plants and animals. They occurred at the end of the Ordovician, Devonian, Permian, Triassic, and Cretaceous periods (see Table 1.6). The most famous of these was at the end of the Cretaceous when the dinosaurs disappeared, as did two-thirds of marine life.

Table 1.7 Major Mass Extinction Events

Time (millions of years ago)	Extinction Event	Description
440	1st Ordovician	Eliminates most marine species; many groups lose more than half their species
360	2nd Devonian	Two-thirds of all species disappear, primarily marine species
251	3rd Permian	More than 90 percent of all species disappear. The worst of the known extinctions
200	4th Triassic	20 percent of species disappear
65	5th Cretaceous-Tertiary (K-T)	The most famous of past extinctions; 60 percent of species died including the dinosaurs
Present day	6th	

Source: Compiled by Laverty.

Each of the first five mass extinctions shown in Table 1.6 and Table 1.7 represents a significant loss of biodiversity—but species diversity recovered over geologic time scales. Mass extinctions are apparently followed by a sudden burst of evolutionary diversification on the part of the remaining species; some evolutionary biologists suggest this is because the surviving species start using habitats and resources that were previously "occupied" by species that went extinct. However, these bursts of diversification do not mean that the recoveries from mass extinction have been rapid; they have usually required some tens of millions of years. The theory describing these brief bursts of evolutionary change is known as *punctuated equilibrium.*

The most striking example of diversification—rapid, on an evolutionary time scale of 10 million years—was the Cambrian explosion. During this relatively short time period, all the animal body plans existing today developed. Before this period, the vast majority of life on earth was restricted to unicellular or undifferentiated multicellular organisms. Scientists still hotly debate what caused such a spectacular spurt of evolutionary radiation and why it subsided, since evolution has proceeded at a more gradual pace subsequently.

Many scientists believe we are on the brink of a "sixth mass extinction," but one that differs from previous events. The five other mass extinctions predated humans, and while theories abound about their causes, they were probably due to some physical process (e.g., catastrophic climate change through meteor impacts or volcanic eruptions), rather than the direct consequence of the action of some other species. In contrast, the sixth mass extinction is the product of human activity over the last several hundred, or even thousand years. In the prehistoric past, humans caused mass extinctions when they migrated to new areas. For example, scientists hypothesize that the large-scale extinction of large mammals (including camels, bears, sloths, saber-toothed tigers, and mammoths) in North America about 12,000 years ago was due to overhunting by humans following colonization. The scale of change in the last 200 years has been similarly severe. Species that were once plentiful have now been drastically reduced in numbers or are completely gone.

The extinction of the passenger pigeon (*Ectopistes migratorius*) in the United States is a dramatic example of human-caused extinction (see Figure 1.4). At the time of the European settlement of North America, passenger pigeons represented 25 to 40 percent of the total bird population of North America. Populations are estimated to have been as high as 3 to 5 billion birds; recent studies suggest that the high population may have also been due to the loss of their predators. Widespread forest clearing and then large-scale hunting of the birds for urban markets eventually led to the passenger pigeon's precipitous decline at the end of the nineteenth century. Once governments and citizens recognized the problem and attempted to address it, however, it was too late. The passenger pigeon required large flocks to breed successfully and could not adapt to survive within smaller flocks. The last known passenger pigeon died in captivity in 1914.

Figure 1.4 The extinction of the passenger pigeon (*Ectopistes migratorius*) in the United States is a dramatic example of human-caused extinction (See Figure 1.6) (*Edouard Poppig 1841*)

It is sometimes difficult to know exactly when a species is extinct in the wild. As they near extinction they are obviously difficult to observe, since they are rare and likely living in remote areas. A species is assumed to be extinct when there is no reasonable doubt that the last individual has died. Usually the rule of thumb is that a species is extinct if after extensive field surveys the species has not been seen for 50 years (though the exact amount of time varies based on the species' lifespan). The Ivory-Billed Woodpecker (*Campephilus principalis*) was thought to be extinct in the 1980s. It may have been rediscovered in 2004, but there is still uncertainty as to whether it has been found again. It is possible the "rediscovery" is simply a misidentified Pileated Woodpecker (*Dryocopus pileatus*).

2

WHERE IS THE WORLD'S BIODIVERSITY?

"...The greatest advance in the knowledge of our planet since the celebrated discoveries of the fifteenth and sixteenth centuries" was how John Murray described the HMS *Challenger* expedition (Murray 1895, xii). Sailing in December 1872, the expedition covered 70,000 miles and laid the foundation for modern oceanography, discovering over 4,700 new marine species.

Historical accounts by naturalists and adventurers exploring new parts of the world emphasize the spectacular diversity of life that exists in the natural world. In today's world, thanks to the likes of the Discovery Channel, the National Geographic, and the Internet, people can easily learn about the diversity of life throughout the world. It's hard for us to comprehend just how foreign and odd the wildlife was for those early explorers in the "new worlds."

This chapter examines the early explorations of the earth and the distribution of life we see on the planet.

During the eighteenth and nineteenth centuries, ships were sent out to survey the world's oceans and map the coastlines (see Table 2.1). Often on these voyages, the ship's surgeon doubled as a naturalist, keeping detailed journals, and collecting unusual plants and animals along the way. These expeditions laid the foundation for modern biology. The most famous of them was that of the HMS *Beagle*. The journey began in December 1831 and lasted for 5 years. The collections made during this journey eventually led to Charles Darwin's formulation of the theories of evolution and natural selection. Darwin's chief scientific rival of sorts—Alfred Russel Wallace—would himself stumble upon the same theory as Darwin (also during a long ship voyage), through his own observations of species distributions on different islands in the Malay Archipelago.

Table 2.1 Major Scientific Expeditions of the Eighteenth and Nineteenth Century

Ship	Captain	Scientist/s	Regions Explored	Dates
Endeavour	James Cook	Joseph Banks, Carl Solander, Herman Sporing	South Pacific, Australia	1768–1771
Resolution	James Cook	Johann Forster, Georg Forster	New Zealand, South Pacific	1772–1775
Bounty	William Bligh	David Nelson	Tahiti, Fiji, and South Pacific	1787–1789
Pizarro		Alexander von Humboldt	South America (Orinoco and Amazon rivers, Andes)	1799–1804
Investigator	Matthew Flinders	Robert Brown	Australia	1801–1805
Beagle	Robert Fitzroy	Charles Darwin	Cape Verde, South America, and South Pacific	1831–1836
Erebus	James Clark Ross	Joseph Hooker	Antarctica, Tasmania, New Zealand, and Kerguelen Island	1839–1843
Rose of Japan		Alfred Russell Wallace	Malay Archipelago [Singapore, Malaysia, Indonesia, and New Guinea]	1854–1862
Vincennes	Charles Wilkes	James Dwight Dana	Western North America, Australia, and South Pacific	1838–1842
Challenger	George Nares	Charles Wyville Thomson	Marine systems (Atlantic, Indian, and Pacific Oceans)	1872–1876

Source: Compiled by Laverty.

∾◦∾

CHALLENGER EXPEDITION

Departing from England in 1872, the HMS *Challenger* would cover 127,000 kilometers (or 68,890 miles) of the world's oceans in its $3^1/_2$-year voyage. The expedition set out to study the natural history and geology of the world's oceans (the first ever to do so). Scientists believed the ocean still had ancient living fossils and hoped the voyage would provide evidence to prove Darwin's theory. With six scientists and a crew of 243 on board, the sheer quantity of data collected was astounding. It took over 19 years to examine the data, and the expedition led to a 50-volume work, 29,552 pages long. The expedition collected 494 depth soundings and 133 dredges. The discoveries included:

- the deepest place in the ocean—the Mariana's trench in the Western Pacific—the spot in the trench where the *Challenger* took a sounding was 35,853 feet (or 10,920 meters) deep and is now known as the "Challenger Deep";
- 4,700 new plants and animals were discovered;
- plots were made of the currents, temperatures, and chemistry of the oceans;
- the outlines of the shapes of the ocean basins were made;
- the first particles that originated in space were discovered (manganese nodules).

NAMING THE WORLD'S BIODIVERSITY

As explorers and scientists began to collect the world's biodiversity, they also began to see the relationships among species and worked to classify them into groups. It was Linnaeus (or Carl von Linné, May 23, 1707–January 10, 1778) a Swedish botanist who established the basis of the scientific naming system still used today. The system he created is outlined in *Systema Naturalae* (1735). Originally published as an 11-page folio, after many revisions it eventually grew to a multivolume edition. Though Linnaeus's original groupings have changed substantially, the basic structure that each species is categorized into a Kingdom, Phylum, Class, Order, Family, Genus, and Species, remains the same. So humans, for example, are classified as follows: Animalia, Chordata, Mammalia, Primates, Hominidae, *Homo sapiens*. The combination of genus and species name is unique to each species; this is known as *binomial nomenclature*.

Linnaeus's classification system centered on the physical characteristics of species, and was initially divided into three kingdoms: plants, animals, and minerals. In 1969 biologists modified this to a system based on five kingdoms: Monera, Protista, Fungi, Plantae, and Animalia. However, this system too is now considered outdated since the discovery of archeabacteria in 1977 by Carl Woese and George E. Fox, and with new understandings of relationships among species based on genetics.

Modern biologists now divide plants and animals into three domains: Archaea, Eubacteria, and Eukaryota. This latter system of classification is centered on the nature of the cells: whether or not they have a nucleus, and the nature of the cell's exterior. However, there is still some dispute over fine-tuning the relationships among the lineages, particularly between the Archaea and the Eukaryota (see Figure 2.1).

Archaea are single-celled species with no nucleus that were originally classified in the kingdom Monera; though still prokaryotes (*organisms*

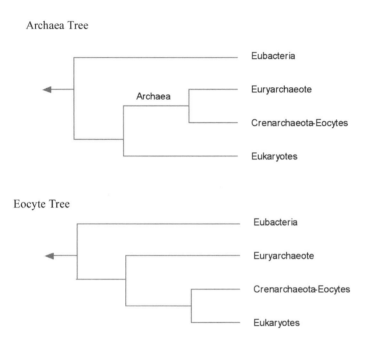

Archaea Tree

Eocyte Tree

Three Major Domains
Eubacteria: true bacteria, mitochondria, and chloroplasts
Eukaryotes: protists, plants, fungi, and animals
Archaea: methanogens, halophiles, sulfolobus, and relatives

Kingdoms in the Archaea Tree or Domains in the Eocyte Tree
Euryarchaeota: methanogens, halophiles, and extreme thermophiles
Crenarchaeota-Eocyste: thermophile and psychrophile and most abundant marine Archea

Note: Extremophiles are species adapted to extreme conditions. Examples include: acidophiles, which live in very acidic environments; halophiles, which live in very salty conditions; thermophiles, which live in very hot temperatures; xerophiles, which live in very dry environments; and psychrophiles orcryophiles, which live in very cold temperatures.

Figure 2.1 Leading Alternative Views of the Major Lineages: Archaea and Eocyte "Trees of Life" (*Perkins*)

that lack a nucleus, or membrane-bound organelles), they are more closely related to eukaroytes than bacteria, as both have similar processes of genetic transcription. Once believed to be found only in extreme habitats, species in the kingdom Archaea are now believed to make up about 20 percent of the diversity in most ecosystems.

Bacteria are also found in all ecosystems. A handful of soil contains about 40 million bacterial cells, a milliliter of freshwater holds a million bacterial cells, and there are about ten times as many bacterial cells as human cells in the human body, most of which live in our digestive tract and on our skin and begin colonizing the human body after we are born. There are 1,000 trillion bacteria cells versus 100 trillion human cells in the human body. Five hundred to a 1,000 species of bacteria live in the average person. The most plentiful bacterium in the world is believed to be *Pelagibacter ubique*; in just 1 milliliter of ocean water there are 250,000 cells. Globally, they weigh more than all the fish in the ocean— their numbers are estimated to reach 24 billion billion billion (or 2.4×10^{28}) cells. They were only discovered in 1990 in a sample from the Sargasso Sea from their rRNA genes, though they are found throughout the world's oceans. In fact, this species has the smallest genome of any self-replicating organism. Little is known about its contribution to ocean ecology and carbon cycling.

Eukarya is made up of organisms whose cells contain a nucleus and a cytoskeleton. This diverse domain includes plants, animals, fungi, slime molds, and many protists.

THE TREE OF LIFE

The affinities of all the beings of the same class have sometimes been represented by a great tree . . . As buds give rise by growth to fresh buds, and these if vigorous, branch out and overtop on all sides many a feebler branch, so by generation I believe it has been with the great Tree of Life, which fills with its dead and broken branches the crust of the earth, and covers the surface with its ever branching and beautiful ramifications. (Charles Darwin, 1866, p. 156)

When the tenth edition of Linnaeus' *Systema Naturae* was published in 1758, he had classified 4,400 species of animals and 7,700 species of plants. Today nearly 2 million species have been identified. But we still have a long way to go. Vertebrates receive the most attention from taxonomists, yet there are about ten times as many plants and a hundred times as many invertebrates. Since Linnaeus we have described about 4,000 to 5,000 insects a year, about 25 percent of which are beetles, and about 400,000 beetles have been described to date. If 2 million remain to be described, at a pace of

3,000 descriptions per year, it will be the year 3056 before we have described all the beetles.

Overall scientific research has shifted from simply naming species to understanding the relationships among them. This effort is still in its infancy as the discovery of genes has offered new ways of understanding the connections among species. Genetic analyses and computers are allowing scientists to further refine and understand the ancestry and relationships among species. Of the nearly 2 million species that have been identified and described, about 80,000 have been placed in different "trees of life"—a metaphor for the relationships among species. The branches of an evolutionary tree, or cladogram, represent relationships among species, while the length of the tree shows the amount of evolution. Biologists around the world are collaborating on an Internet version of the "tree of life," with over 9,000 web pages, which explores the diversity and evolutionary history among species (see http://www.tolweb.org/tree/).

Patterns of Diversity

Biogeography

Biogeography is the study of how species are distributed across space and through time. Analyses of these patterns of biological diversity can be divided into two scientific disciplines: historical biogeography and ecological biogeography

Historical biogeography examines past geological events in Earth's history and uses these to explain patterns in the spatial and temporal distribution of organisms (usually species or higher taxonomic ranks). For example, an explanation of the distribution of closely related groups of organisms in Africa and South America is based on the understanding that these two landmasses were once connected as part of a single landmass (known as Gondwana). The ancestors of those related species, which are now found in Africa and South America, are assumed to have had a cosmopolitan distribution stretching across both continents when they were connected. Following the separation of the continents by the process of plate tectonics, the isolated populations are assumed to have undergone allopatric speciation (i.e., speciation achieved between populations that are completely geographically separate). This separation resulted in the closely related groups of species on the now separate continents. Clearly, an understanding of the systematics of the groups of organisms (i.e., the evolutionary relationships that exist between the species) is an integral part of these historic biogeographic analyses. The world's major biogeographic realms include the: Nearctic, Neotropic, Antarctica, Palearctic, Afrotropic, IndoMalay, Australasia, and Oceania

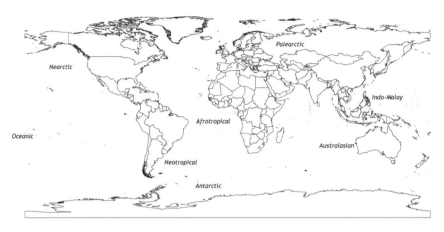

Figure 2.2 Map of the World's Major Biogeographic Realms (*Horning/ Laverty © CBC-AMNH*. Used by permission)

(see Figure 2.2). These large regions have a broadly similar biological evolutionary history.

The same historical biogeographic hypotheses can be applied to the spatial and temporal distributions of marine biota. For example, the biogeography of fish from different ocean basins is associated with the geological evolution of these basins. However, we cannot assume that all existing distribution patterns are solely the product of these past geological processes. For example, the current marine life of the Mediterranean can be attributed to the basin's complex geological history as well as to dispersal events. The Mediterranean separated from the Indian and Atlantic Oceans, underwent periods of extensive desiccation followed by flooding and recolonization by species from the Atlantic. Geology is not the only force that shaped current diversity in the Mediterranean; the eastern Mediterranean has been colonized more recently by species that have dispersed from the Red Sea via the Suez Canal, which was completed in 1869.

The field of *ecological biogeography* focuses on current populations and interactions of species on shorter time scales. It first examines the dispersal of organisms (usually individuals or populations) and the mechanisms that influence this dispersal, and then uses this information to explain the spatial distribution patterns of these organisms. Ecological biogeography often considers the number of species an area supports, or its species richness. It was the theory of island biogeography that changed biogeography from a primarily historical focus. The term, island biogeography, was first coined in the 1960s by Robert MacArthur and E.O. Wilson. According to the theory the number of species on an island can be predicted based on the distance from the mainland and the size of the island.

--- ✐ ---

WALLACE'S LINE

While traveling from the island of Bali to Lombok, in what is now Indonesia, Alfred Russel Wallace stumbled upon something unusual. This discovery came during his 8-year expedition exploring the Malay Peninsula that began in 1852 and its associated islands. Though the island of Lombok was just 35 kilometers east of Bali, Wallace was surprised to discover a completely different set of birds. Rather than resembling the Asiatic bird species of Bali, they looked more like Australian birds. Other species with Asian origins (including tigers, rhinoceros, orangutans, and the bird groups, the barbets and the trogons), are only found west of the "line," while to the east were friarbirds, cockatoos, birds of paradise, as well as various marsupials, all apparently descended from Australian species. The line that marked the boundary between species that fall into Asia and Australasia was eventually known as "Wallace's Line."

While Wallace noticed the unusual distribution of species on the islands he did not recognize the reason for the difference. Islands to the east of the line are all part of the Sunda Shelf, so that when sea levels were lower the islands of Borneo, Java, Sumatra, and Bali were all once connected to Mainland Southeast Asia (see Figure 2.3), whereas beneath the sea a deep-water trench separates Bali and Lombok, preventing connectivity even during low sea levels.

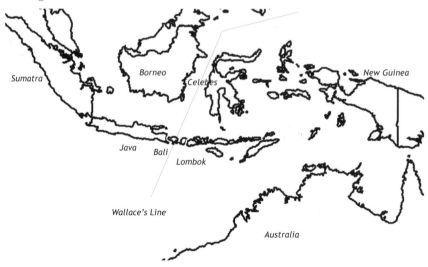

Figure 2.3 Map of Indonesia depicting Wallace's Line (*Horning/Laverty* © *CBC-AMNH*. Used by permission)

--- ✐ ---

Islands Not Evolutionary Dead Ends

It was once thought that continents provided the source for species found on islands, and the process was one-way, with continents feeding islands with species. Thus, the further an island was from a continent the fewer species it had. A 2005 study by ornithologists Christopher Filardi and Robert Moyle turned this traditional vision of species distribution and evolution on its head. Their study examined the genetic ancestry of a group of South Pacific island birds, the Monarch flycatchers, which revealed that some species in fact recolonized Australia and New Guinea. Thus, islands can lead to speciation and these species can later repopulate continents.

Why Are There More Species in the Tropics?

Generally speaking, there are more species in warm tropical ecosystems than cold temperate ecosystems found at high latitudes. In fact, not only are there more species in the tropics, but there are also more groups at higher taxonomic levels (i.e., genera, families). This distribution applies for many different groups from mammals to insects. Various hypotheses have been raised to explain these patterns of species richness; however, the debate is ongoing as to the exact cause as the hypotheses are difficult to test.

One of the leading theories, often called the "climate-speciation hypothesis," links warmer temperatures to higher rates of speciation. According to one branch of the theory there is a direct link between the warmer climate of the tropics and increased speciation. Warmer temperatures lead to increased mutation rates, and thus increased molecular evolution and more species. A second branch of this theory indirectly connects a warmer climate to increased growth rates and thus shorter generation times, at which natural selection operates, and thus leads to increased speciation rates. Alternatively shorter generation times could also lead to faster molecular evolution and again increased speciation rates.

Another theory is based on the assumption that warm, moist, tropical environments, with long day-lengths and abundant water provide organisms with more resources for growth and reproduction than harsh environments with few energy resources. When environmental conditions favor the growth and reproduction of primary producers, such as plants, aquatic algae, and corals, then these may in turn support large numbers of secondary consumers, such as small herbivores, which also support a more numerous and diverse fauna of predators. In contrast, the development of primary producers in colder, temperate ecosystems is constrained by seasonal changes in sunlight and temperature. Consequently, these temperate ecosystems support fewer secondary consumers and predators.

Species and ecosystem richness also varies with altitude—typically mid-elevations have the highest species diversity, while the highest elevations have fewer species. Mountainous environments can be subdivided vertically into altitudinal belts, each with a different ecosystem. Climatic conditions at higher elevations usually have lower temperatures and humidity, conditions under which fewer species can survive. In oceans and freshwater ecosystems, the pattern reverses itself; with increasing depth below the surface, species richness declines. However, in the oceans there may be a rise in species richness close to the seabed, which is associated with an increase in ecosystem heterogeneity.

SETTING CONSERVATION PRIORITIES: WHERE TO CONSERVE?

By mapping spatial gradients in biodiversity, we can also identify areas of special conservation interest. Extrapolating from existing data sets on scientifically surveyed areas of the world, it is possible to map out where different species and ecosystems are found, where the greatest species richness is located, and where rare species and ecosystems are found. Understanding species and ecosystem distributions aid conservation biologists in selecting regions to conserve. For example, conservation biologists are often interested in areas that have a high proportion of *endemic species*, or species whose distributions are naturally restricted to a limited area. Areas that support a large number of endemic species are obvious choices for conservation, since those species are found nowhere else and if they disappear, these species are lost forever (see Figure 2.4). Conservation biologists may also seek to conserve areas that are particularly threatened or that host a large number of species. Sometimes areas of high endemism overlap with areas of high species richness; however, this varies greatly with the taxa and region. In fact, a recent global study of birds found that biological hotspots of endemism, species richness, and threat, only overlapped in 2.5 percent of cases. Two key approaches for deciding what and how to conserve biodiversity are the ecoregion and the hotspot approach outlined below.

Ecoregions

Both the World Wide Fund for Nature (WWF) and the Nature Conservancy (TNC) use the ecoregion approach to conservation planning. Scientists at the WWF initially identified a comprehensive list of 825 terrestrial and 450 freshwater ecoregions on the planet. Among these

Figure 2.4 Madagascar is famous for its high level of endemism—many species found there are found nowhere else in the world. Among these unique species are lemurs Ring-tailed lemur (*Frey©CBC-AMNH*. Used by permission)

ecoregions, they then selected the top priorities for global conservation called—the "*Global 200*"—though these actually cover 238 ecoregions: 142 terrestrial, 53 freshwater, and 43 marine ecoregions (see Table 2.2). These ecoregions were selected because they harbor exceptional biodiversity and are representative of the variety of Earth's ecosystems. They considered factors such as species endemism, richness, and global rarity. Each ecoregion is also assigned a conservation status: critical or endangered, vulnerable, stable or intact. Over half of the *Global 200* are endangered. One outcome of this approach was that some ecosystems, such as Mediterranean forests, were found to be more endangered than tropical ones.

Also, in 2007 TNC and WWF launched a parallel program focused solely on the marine realm—the first ever classification of the world's coastal areas. The Marine Ecosystems of the World (MEOW) centers on identifying key areas of the coast and shelf, classifying them into 12 realms, 62 provinces, and 232 ecoregions.

Biodiversity Hotspots

Conservation International's (CI) approach to conservation priority setting centers on protecting regions called *biodiversity hotspots* (see

Table 2.2 The "Global 200" Ecoregions Identified By the World Wide Fund for Nature (WWF) As Conservation Priorities

Tropical and Subtropical Moist Broadleaf Forests

Afrotropic
  Guinean moist forests
  Congolian coastal forests
  Cameroon Highlands forests
  Northeastern Congolian lowland forests
  Central Congolian lowland forests
  Western Congolian moist forests
  Albertine Rift montane forests
  East African Coastal Forests
  Eastern African montane forests
  Madagascar lowlands and subhumid forests
  Seychelles and Mascarene Islands moist forests
Australasia
  Sulawesi moist forests
  Moluccas moist forests
  Southern New Guinea lowland forests
  New Guinea montane forests
  Solomons-Vanuatu-Bismarck moist forests
  Queensland tropical rain forests
  New Caledonia moist forests
  Lord Howe-Norfolk Islands forests
Indomalaya
  South Western Ghats montane rain forests and moist deciduous forests
  Sri Lanka moist forests
  Northern Indochina Subtropical moist forests
  Southeast China-Hainan moist forests
  Taiwan montane forests
  Annamite Range moist forests
  Sumatran Islands lowland and montane forests
  Philippines moist forests
  Palawan moist forests
  Kayah-Karen/Tenasserim moist forests
  Peninsular Malaysian lowland and montane forests
  Borneo lowland and montane forests
  Nansei Shoto Archipelago forests
  Eastern Deccan Plateau moist forests
  Naga-Manipuri-Chin hills moist forests
  Cardamom Mountains moist forests
  Western Java montane forests
Neotropic
  Greater Antillean moist forests
  Talamancan-Isthmian Pacific forests
  Choco-Darien moist forests
  Northern Andean montane forests

Table 2.2 (*Continued*)

Coastal Venezuela montane forests
Guianan moist forests
Napo moist forests
Rio Negro-Jurua moist forests
Guyana Highlands moist forests
Central Andean yungas
Southwestern Amazonian moist forests
Atlantic forests
Oceania
South Pacific Islands forests
Hawaii moist forests

Tropical and Subtropical Dry Broadleaf Forests

Afrotropic
Madagascar dry deciduous forests
Australasia
Nusa Tenggara dry forests
New Caledonia dry forests
Indomalaya
Indochina dry forests
Chhota-Nagpur dry forests
Neotropic
Mexican dry forests
Tumbesian-Andean valleys dry forests
Chiquitano dry forests
Atlantic dry forests
Oceania
Hawaii dry forests

Tropical and Subtropical Coniferous Forests

Nearctic
Sierra Madre Oriental and Occidental pine-oak forests
Neotropic
Greater Antillean pine forests
Mesoamerican pine-oak forests

Temperate Broadleaf and Mixed Forests

Australasia
Eastern Australia temperate forests
Tasmanian temperate rain forests
New Zealand temperate forests
Indomalaya
Eastern Himalayan broadleaf and conifer forests
Western Himalayan temperate forests

(*Continued*)

Table 2.2. (*Continued*)

Nearctic
    Appalachian and mixed mesophytic forests
Palearctic
    Southwest China temperate forests
    Russian Far East temperate forests

Temperate Coniferous Forests

Nearctic
    Pacific temperate rain forests
    Klamath-Siskiyou forests
    Sierra Nevada forests
    Southeastern coniferous and broadleaf forests
Neotropic
    Valdivian temperate rain forests-Juan Fernandez Islands
Palearctic
    European-Mediterranean montane mixed forests
    Caucasus-Anatolian-Hycanian temperate forests
    Altai-Sayan montane forests
    Hengduan Shan coniferous forests

Boreal Forests/taiga

Nearctic
    Muskwa-Slave Lake boreal forests
    Canadian taiga
Palearctic
    Ural Mountains taiga
    East Siberian taiga;
    Kamchatka taiga and grasslands

Tropical And Subtropical Grasslands, Savannas, and Shrublands

Afrotropic
    Horn of Africa acacia savannas
    East African acacia savannas
    Central and Eastern miombo woodlands
    Sudanian savannas
Australasia
    Northern Australia and Trans-Fly savannas
Indomalaya
    Terai-Duar savannas and grasslands
Neotropic
    Llanos savannas
    Cerrado woodlands and savannas

Temperate Grasslands, Savannas, and Shrublands

Nearctic
    Northern prairie

Table 2.2. (*Continued*)

Neotropic
  Patagonian steppe
Palearctic
  Daurian steppe

Flooded Grasslands and Savannas

Afrotropic
  Sudd-Sahelian flooded grasslands and savannas
  Zambezian flooded savannas
Indomalaya
  Rann of Kutch flooded grasslands
Neotropic
  Everglades flooded grasslands
  Pantanal flooded savannas

Montane Grasslands and Shrublands

Afrotropic
  Ethiopian Highlands
  Southern Rift montane woodlands
  East African moorlands
  Drakenberg montane shrublands and woodlands
Australasia
  Central Range subalpine grasslands
Indomalaya
  Kinabalu montane shrublands
Neotropic
  Northern Andean paramo
  Central Andean dry puna
Palearctic
  Tibetan Plateau steppe
  Middle Asian montane steppe and woodlands
  Eastern Himalayn alpine meadows

Tundra

Nearctic
  Alaskan North Slope coastal tundra
  Canadian low arctic tundra
Palearctic
  Fenno-Scandia alpine tundra and taiga
  Taimyr and Russian coastal tundra
  Chukote coastal tundra

Mediterranean Forests, Woodlands, and Shrub

Afrotropic
  Fynbos

(*Continued*)

Table 2.2. (*Continue*)

Australasia
    Southwestern Australia forests and scrub
    Southern Australia mallee and woodlands
Nearctic
    California chaparral and woodlands
Neotropic
    Chilean Matorral
Palearctic
    Mediterranean forests, woodlands, and shrub

Deserts and Xeric Shrublands

Afrotropic
    Namib-Karoo-kaokoveld deserts
    Madagascar spiny thicket
    Socotra Island desert
    Arabian Highland woodlands and shrublands

Australasia
    Carnavon xeric scrub
    Great Sandy-Tanami deserts
Nearctic
    Sonoran-Baja deserts
    Chihuahuan-Tehuacan deserts
Neotropic
    Galapagos Islands scrub
    Atacama-Sechura deserts
Palearctic
    Central Asian deserts

Mangroves

Afrotropic
    Gulf of Guinea mangroves
    East African mangroves
    Madagascar mangroves
Australasia
    New Guinea mangroves
Indomalaya
    Sundarbans mangroves
    Greater Sundas mangroves
Neotropic
    Guianan-Amazon mangroves
    Panama Bight mangroves

Source: World Wide Fund for Nature.

Table 2.3 The World's Biodiversity Hotspots Identified By Conservation International. Hotspots Are Selected Based on Species Richness, Endemism, and Degree of Threat That They Face

North and Central America
    Caribbean Islands
    California Floristic Province
    Madrean Pine-Oak Woodlands
    Mesoamerica
South America
    Atlantic Forest
    Cerrado
    Chilean Winter Rainfall-Valdivian Forests
    Tumbes-Chocó-Magdalena
    Tropical Andes
Europe and Central Asia
    Caucasus
    Irano-Anatolian
    Mediterranean Basin
    Mountains of Central Asia
Africa
    Cape Floristic Region
    Coastal Forests of Eastern Africa
    Eastern Afromontane
    Guinean Forests of West Africa
    Horn of Africa
    Madagascar and the Indian Ocean Islands
    Maputaland-Pondoland-Albany
    Succulent Karoo
Asia-Pacific
    East Melanesian Islands
    Himalaya
    Indo-Burma
    Japan
    Mountains of Southwest China
    New Caledonia
    New Zealand
    Philippines
    Polynesia-Micronesia
    Southwest Australia
    Sundaland
    Wallacea
    Western Ghats and Sri Lanka

Source: Conservation International.

Table 2.3). Norman Myers first identified ten tropical forest *hotspots* based on plant endemism and threat in 1988, and his method was later adopted by CI in 1989. The method of selecting a hotspot has been refined since then. A *terrestrial biodiversity hotspot* is now defined quantitatively as an area that has at least 0.5 percent, or 1,500 of the world's 300,000 species of green plants, and that has lost at least 70 percent of its primary vegetation. *Marine biodiversity hotspots* are quantitatively defined based on measurements of relative endemism of multiple taxa (i.e., species of corals, snails, lobsters, and fish) within a region and the relative level of threat to that region. According to this approach, the Philippine archipelago and the islands of Bioko, São Tomé, Principe, and Annobon in the eastern Atlantic Gulf of Guinea are ranked as two of the most threatened marine biodiversity hotspots. The CI hotspot approach has continued to evolve, for example, boundaries have been updated and streamlined to conform with the WWF/TNC Ecoregion approach. Today CI recognizes thirty-four hotspots; these hotspots once covered 15.7 percent of the planet, but already 86 percent of the hotspots have been destroyed and they now cover just 2.3 percent of the planet.

To complement its hotspot program, CI has also identified five key remaining Wilderness Areas: Amazonia, Congo Basin, New Guinea, North American deserts, and Southern Africa. In contrast to the hotspots, these areas are largely intact with more than 70 percent of their original vegetation; however, they each harbor significant portions of the planet's biodiversity.

Living Landscapes

The Wildlife Conservation Society has developed two strategies for conservation priority setting: Living Landscapes and the Last of the Wild. The Living Landscapes program is a wildlife-based conservation strategy. The focus is on developing practical site-based approaches to conserving wildlife. The Last of the Wild are 568 areas, which are the largest and wildest places left within their biome. The areas are assessed for "wildness" by looking at various measures of human influence, including population and road density, and land use patterns.

A set of seventeen "megadiverse" countries has also been identified by UNEP's World Conservation Monitoring Centre (WCMC); at the top of the list are Australia, Brazil, China, Columbia, and the Democratic Republic of Congo.

DEEP-SEA DIVERSITY

It was long thought that all life on earth ultimately depended on photosynthesis, the process by which plants convert sunlight to sugars. It wasn't until explorations of the deep ocean in the 1970s that this myth was laid to rest. Exploring the ocean bottoms using remotely guided cameras, scientists uncovered whole communities of strange animals never seen before. These communities centered on hydrothermal vents. Hydrothermal vents are found deep in the world's oceans in areas where the earth is volcanically active, such as where tectonic plates are separating, Found at depths of about 2,000 meters, the vents, also known as black smokers, spew out "clouds" of smoke filled with dissolved minerals from deep within the earth. The smoke might extend up 45 feet, and surrounding waters are very acidic and have temperatures up to 600 degrees F.

Life at these vents is supported by chemosynthesis—a process whereby bacteria convert carbon molecules into organic matter using hydrogen sulphide, hydrogen, or methane. Among the amazing and unusual creatures living in these deep-water vents are giant tube worms (*Riftia pachyptila*). Growing up to 8 feet (2.4 meters) long, the worms have no mouth or digestive track, instead absorbing nutrients through a red feathery plume at the tip. The plume is filled with hemoglobin unlike that found in other species as it can carry oxygen even when sulphide is present.

Conservation biologists are also interested in areas that have relatively low biological diversity but also include threatened or rare species (sometimes called *biodiversity coldspots*). Just has hotspots do not imply that the ecosystem is physically "hot" (although most hotspots are coincidentally located in the hot tropics), coldspots similarly are not necessarily "cold." Although these areas are low in species richness, they can also be important to conserve, as an individual "coldspot" may be the only location where a rare species is found. Extreme physical environments (low or high temperatures or pressures, or unusual chemical composition) inhabited by just one or two specially adapted species are "coldspots" that warrant conservation because they represent unique environments that are biologically and physically interesting.

3

WHY IS BIODIVERSITY IMPORTANT?

The last word in ignorance is the man who says of an animal or plant: "What good is it?"

(Aldo Leopold, *Round River*, 1993, pp. 145–146)

People depend on biodiversity for so many things, from food to medicine, yet few appreciate its value. This chapter examines why biodiversity is important, both in terms of the direct goods it supplies to people, such as food and timber, as well as the less tangible values, such as aesthetic or inspirational values.

INTRODUCTION

Humans depend upon biodiversity in many ways: to satisfy basic needs like food and medicine, and to enrich our lives culturally or spiritually. Yet in an increasingly modern, technological world, people often forget how fundamental biodiversity is to their daily life and are unaware of the impact of its loss.

Despite its importance, determining the value or worth of biodiversity is complex and often a cause for debate. This is largely due to the fact that the worth placed on biodiversity is a reflection of underlying human values, and these values vary dramatically among societies and individuals. The perspective of rural versus urban dwellers toward wildlife is one example. People in cities appreciate elephants for their sheer size, charisma, and intelligence. People in rural areas who live near elephants; however, tend to perceive them as a threat to people and their crops and property.

Values are dynamic; they change over time and vary according to the situation. While I might think a bear cub is cute when I see it at the zoo, I'm certain to feel differently when I meet one while hiking. Both the diversity of values toward a species and the changes in values over time can be examined in the case of the gray wolf in the United States. Once widespread throughout North America, the gray wolf (*Canis lupus*) was deemed a threat to human livelihoods and was systematically hunted, beginning in the 1600s. By the late 1800s, it had been virtually exterminated from its original range. But in the late 1970s, values shifted in response to dwindling wolf populations and the U.S. Government began programs to restore them to their former range. To some Americans, wolves have come to signify the "wilderness" and are an important tourist attraction in Yellowstone National Park and northern Minnesota. For others, the image of the wolf as a threat and a nuisance persists to this day. The debate over wolf restoration programs demonstrates not only the changes in values but also the multiplicity of values within any one society.

CATEGORIZING BIODIVERSITY'S VALUE

There are many ways to categorize the value of biodiversity; typically they can be divided into:

- Utilitarian (also known as instrumental, extrinsic, or use) values, and
- Intrinsic (also known as inherent) values.

A living thing's *utilitarian value* is determined by its use or function. Usually utilitarian value is measured in terms of its use for humans, such as for medicine or food. However, it can also represent the value of an organism to other living things or its ecological value; pollinators, such as bees, are essential to the reproduction of many plants. In contrast, *intrinsic value* describes the inherent worth of an organism, independent of its value to anyone or anything else. In other words, all living things have a right to exist—regardless of their utilitarian value.

Utilitarian Values

Determining the value or worth of biodiversity is complex. Economists typically subdivide utilitarian or use values of biodiversity into *direct use value* for those goods that are consumed directly, such as food, and *indirect use value* for those services that support the items that are consumed, including ecosystem functions such as nutrient cycling.

Alternatively, the United Nations Millennium Ecosystem Assessment grouped all of these values under *ecosystem services* and then further subdivided these as *provisioning services* (food, fiber, fuel, genetic resources, medicines, pharmaceuticals and other chemicals, ornamental resources); *regulating services* (air quality, climate, water, erosion, disease, pest and natural hazard regulation; pollination); *cultural services* (cultural, spiritual, knowledge systems, educational, inspiration, aesthetic, social relations, sense of place, cultural heritage values, recreation, and ecotourism); and *supporting services* (soil formation, photosynthesis, primary production, nutrient cycling).

There are several less tangible values that are sometimes called *nonuse or passive values*, for things that we don't use but we would consider as a loss if they were to disappear; these include *existence value*, the value of knowing something exists even if you will never use it or see it, and *bequest value*, the value of knowing something will be there for future generations. *Potential or Option value* refers to the use that something may have in the future; sometimes this is included as a use value; we have chosen to include it within the passive values here based on its abstract nature.

Scientists vary in which components they include within the category of "utilitarian" values. For example, some classify spiritual, cultural, and aesthetic values as indirect use values, that is, they provide an indirect service by enriching our lives, while others consider them to be nonuse values, differentiated from indirect use values—such as nutrient cycling—because spiritual, cultural, and aesthetic values are not essential to human survival. Still others consider these values as separate categories entirely. Regardless of how they are categorized, what is key is that there are many ways to value biodiversity (see Table 3.1).

Table 3.1 Ways We Value Biodiversity

Direct Use Value (Goods or Provisioning Services)	Indirect Use Value (Ecosystem, Regulating, Cultural, and Supporting Services)	Other Values
Food, medicine, building material, fiber, fuel	Atmospheric and climate regulation, pollination, nutrient recycling; Cultural, Spiritual, and Aesthetic*	Potential (or Option) Value[1]; Existence Value[2]; Bequest Value[3]

* Some authors choose to differentiate these values from those services that provide basic survival needs, such as the air we breathe, referring to them as Cultural Services, and Regulating or Supporting Services respectively.

[1] Potential Value—Future value either as a good or service.

[2] Existence Value—Value of knowing something exists.

[3] Bequest Value—Value of knowing that something will be there for future generations.

Source: Compiled by Laverty.

Direct Use Values: Goods or Provisioning Services

The earth provides an abundance of goods essential to human life—so many, indeed, that it is difficult to create a comprehensive list of them. Some examples include food, shelter, timber, fuel, clothing, fiber, industrial products, and medicine.

Food

Humans have spent most of their roughly 200,000-year existence as hunter-gatherers, dependent on wild plants and animals for survival. Around 10,000 years ago, the first plants were cultivated, marking a fundamental shift in human history. Biodiversity played a central role in the development of agriculture, providing the original source of all crops and domesticated animals. And today people still depend on biodiversity to maintain healthy, sustainable agricultural systems.

Humans have used over 12,000 wild plants for food, yet now only twenty species support much of the world's population. Of all the plants that we depend on, none are more important than the grass family, the Gramineae. The grass family includes the world's principal staples: wheat (*Triticum aestivum*), rice (*Oryza sativa L.*), and corn (maize) (*Zea mays*). Rice and corn formed the basis of civilizations in the Far East and the Americas, while wheat, together with barley, formed the basis of the civilizations in the Near East.

Though agriculture depends on relatively few plants and animals, genetic diversity is essential to improve the productivity of crops and livestock, and to create varieties and breeds that are resistant to pests or disease. Wild species still provide an important source of genetic diversity. With the expansion of trade, it is easy to forget that at one time domestic species were once wild plants and only domesticated by peoples who lived where the wild species was found. It is hard to imagine Italian cuisine without tomatoes; yet tomatoes, originally from South America, only spread around the world after the Spanish brought the plant to Europe and Asia in the 1500s. A similar story could be told for many other domesticated plants; a detailed table of the centers of origins of the world's major plant species reminds us that once apples were only found in Central Asia and oranges only in Southeast Asia, India, and China (see Table 3.2).

Biodiversity acts as a form of insurance for agriculture by helping to ensure that crops can adapt to future environments. Changing climates may require drought-resistant or salt-tolerant crops, for instance. Sorghum, emmer, and spelt were once widely grown grains but have

Table 3.2 In Our Modern World, Where Food Is Grown and
Transported Far from Where It Originated, It Is Easy to Forget Where
Domesticated Plants Originated.
(Vavilov first identified the eight major centers of crop origin in the late
1930s. Below are a selection of crops and their centers of origin.)

I. A. East Asia
  a. Buckweat
  b. Peach
  c. Soybean
  d. Sweet orange
  e. Tea
  B. Southeast Asia
  f. Banana
  g. Clove
  h. Coconut palm
  i. Cucumber
  j. Eggplant
  k. Tangerine
  l. Sugarcane
II. India/Indo-Malaya
  a. Black pepper
  b. Chickpea
  c. Lime
  d. Mango
  e. Orange
  f. Rice
  g. Sesame
III. Central Asia (North India, Afghanistan, Turkmenistan)
  a. Almond
  b. Apple
  c. Carrot
  d. Garlic
  e. Onion
  f. Pear
  g. Spinach
IV. Near East (Transcaucasia, Iran, highlands of Turkmenistan)
  a. Cherry
  b. Durum wheat
  c. Einkorn wheat
  d. Lentil
  e. Plum
  f. Rye
V. Mediterranean
  a. Asparagus
  b. Cabbage

(*Continued*)

Table 3.2 (*Continued*)

---

   c. Cauliflower
   d. Celery
   e. Durum wheat
   f. Olive
   g. Sugar beet
VI. Horn of Africa (Djibouti, Eritrea, Ethiopia, Somalia)
   a. Coffee
   b. Emmer
   c. Sorghum
   d. Pearl Millet
   e. Okra
   f. Yam
VII. Southern Mexico and Central America
   a. Amaranth
   b. Avocado
   c. Corn
   d. Lima bean
   e. Pepper
   f. Squash
   g. Sweet potato
   h. Tomato
   i. Vanilla
VIII. South America (Northeastern South America, Bolivia, Ecuador, Peru, Chile)
   a. Cacao or Cocoa
   b. Peanut
   c. Pepper
   d. Potato
   e. Pineapple
   f. Tomato

Source: Compiled by Laverty, adapted from Vavilov

been largely replaced by wheat. However, because of their unique environmental adaptations—sorghum, for example, can be grown in drier climates that do not support wheat—these grasses may become more important in the future, should climatic conditions change.

For many rural peoples in developing countries, wild species are still an important source of food and income, including green leafy plants, fruits, fungi, nuts, and meat. Furthermore, with the exception of only a few species, the world's marine fisheries are dominated by wild-caught fish, representing 82.4 percent of the 104.1 million tons produced in 2004, according to the Food and Agriculture Organization (FAO).

Less-familiar wild plants may become important foods in the future. For example, Peach Palm (*Guilielma gasipaes or Bactris gasipaes* Kunth, Arecaceae), also know as Pejibaye, from Central America, produces one of the most balanced foods for human nutrition—an ideal mixture of carbohydrates, protein, fat, vitamins, and minerals. Currently Peach Palm is cultivated for the heart of palm and its fruits, especially in Costa Rica and Brazil. Peach Palm plantations produce more protein and carbohydrate per hectare than corn.

Some "forgotten" species are also being revived. Amaranth, first domesticated in Central America over 6,000 years ago, was an important food crop for both the Inkas of Peru and Mexico's Aztecs. Its production declined in the early 1500s after Spanish conquerors discouraged rituals and objects associated with Aztec culture and religion.

Related to beets and spinach, amaranth is edible as both a vegetable and a grain. The vitamin-packed grain is higher in protein than beans, has more fiber than wheat or soybeans, and is exceptionally rich in the amino acid lysine (rare in the plant world). Amaranth also contains more calcium and magnesium than milk and four times the iron of brown rice. Because of its high nutritional value, amaranth is being used to combat malnutrition in many parts of the world.

Though most species of plants and animals farmed today were domesticated between 2,000 and 10,000 years ago, since 1900 many aquatic species have been domesticated. About 250 marine and 180 freshwater species are now farmed, including crustaceans, echinoderms, mollusks, worms, algae, and fish. Aquaculture production continues to expand rapidly.

Building Materials, Paper Products, and Fuel Wood

Trees and several grasses, most notably bamboo and rattan, are basic commodities used worldwide for building materials, paper products, and fuel. The worldwide production of timber and related products—including homes, furniture, mulch, chipboard, paper and packaging—is a multibillion-dollar industry.

The common name "bamboo" describes many different species and genera. All of these are part of the subfamily Bambusoidaeae (of the family Poaceae, or Gramineae, the grass family), which comprises both woody and herbaceous bamboos and includes about 1,575 species. According to the International Network of Bamboo and Rattan, over 1 billion people use bamboo for housing. Given that it is lightweight and strong, and can be grown sustainably (grows fast and usually without pesticides), bamboo is being used for an increasing range of products, from furniture to fabric. Bamboo is even used instead of steel rods in large skyscrapers in Asia.

Rattan (the common name describes 13 genera and about 700 species of the subfamily Lepidocaryoide) though similar looking it is not as widely used as bamboo. Its primary use is for cane furniture, as well as for matting, basketry, and handicrafts.

Though easy to forget in a developed country, one of the most important uses of wood is for fuel. Wood was the main fuel everywhere in the world until the middle of the nineteenth century, after which it began to be replaced by other energy sources, such as oil, in industrialized countries. Fuel wood, charcoal, and other fuel derived from wood are still the major source of energy in low-income countries; the major consumers and producers of wood for fuel are Brazil, China, India, Indonesia, and Nigeria. According to the World Resources Institute (based on FAO 2007), about 51 percent of all harvested wood is used as fuel, burned either directly or after being converted to charcoal (see Figure 3.1); while in some developing countries this figure can be as high as 90 percent.

Some of the earliest papers were made in Egypt about 2,000 years ago from papyrus (*Cyperus papyrus*), a reed that grows in wetlands along the Nile River. To make paper, the inner part of the papyrus stalk was cut into strips and soaked in water. Two layers of the strips were pressed together, held by the plant's natural glue. A number of different plants have been used to make paper, including rice and bamboo. Today paper is made from wood pulp or from a number of different crops, primarily wheat, rice, and sugarcane (in Europe and North America the industry depends on wood while in Asia they use agricultural crops). To make paper from wood, it is first chipped and then combined with water and chemicals to produce a slurry or pulp. The pulp is then sprayed onto a thin wire, and eventually pressed and dried. Paper can also be made from the by-product of the harvest of some agricultural crops, and it is a more sustainable alternative to wood. For example, when wheat, rye, flax, corn, and other straws are harvested, much of the vegetation from crop is simply wasted (such as the corn stalk, etc.); each year about 200 million tons is simply burned after the harvest. This agricultural waste can be put to use and converted to paper. The size of the fibers vary, for example wheat straw produces a small fiber that can be combined with a longer fiber to enhance the quality of the paper.

Industrial Products

Many industrial products are extracted from plants and animals. Some of the most important of these are cork, rubber, latex, shellac, resins,

Figure 3.1 Trees and grasses, like bamboo and
rattan, are harvested for fuel and building
particularly in rural areas of the developing world
(*Frey©CBC-AMNH.* Used by permission)

perfumes, waxes, and oils. A list of a few of these products and their
source is provided in Table 3.3.

In addition to fuel oil, many common chemical products like plastics,
fertilizers, and pesticides are manufactured from petroleum. Petroleum
and other fossil fuels (coal and natural gas) are a product of millions of
years of geologic processes acting on the remains of plants and animals
trapped under layers of sediment.

Table 3.3 A Few Industrial Products Extracted from Plants and Animals

Originating Plant Or Animal	Product/End Use
Cork oak (*Quercus suber*)	Cork
Pará rubber tree (*Hevea brasiliensis*)	Rubber
Lac insect (*Laccifer spp.*)	Shellac
Carnauba palm (*Copernicia cerifera*)	Carnauba wax
Wax plant (*Euphorbia antisyphilitica*)	Candelilla wax
Jojoba plant (*Simmondsia chinensis*)	Jojoba oil
Cochineal insect (*Dactylopius coccus*)	Carmine dye*

Note: *Carmine is the only red dye approved by the U.S. Food and Drug Administration for use in foods, drugs, and cosmetics.
Source: Compiled by Laverty.

Endangered Cork Forests

Cork forests, once found throughout large parts of the Mediterranean, now cover 2.7 million hectares of Portugal, Spain, Algeria, Morocco, Italy, Tunisia, and France. Unlike other Mediterranean habitats, cork oak forests were never completely cleared in the western Mediterranean and have some of the richest biodiversity in the region.

Cork oak (*Querus suber*) is one of 150 endemic trees in these landscapes. Many cork oak landscapes are interspersed with pasturelands and crops; these model landscapes thus not only conserve biodiversity but also sustain local, rural livelihoods. Levels of plant diversity in these forests can reach 135 species per square meter. These landscapes are the breeding grounds of more than 100 songbirds and over 160 bird species use the habitat. They are also crucial habitat for a number of endangered species including the Iberian Imperial Eagle (*Aquila adalberti*) (ca. 150 remain), the Iberian lynx (*Lynx pardinus*) (ca. 100 remain), and the Barbary deer (*Cervus elaphus barbarus*) (ca. 200 remain).

Cork is harvested from living trees by peeling off the outer bark or "cork cambium." The tree survives and can keep producing cork for up to 150 years. This sustainable resource has a slow start. The first cork can only be harvested when the tree is 25 years old. Subsequently, it can only be harvested about once in a decade. The initial cork harvested is of a lower quality only suitable for flooring and only after a tree reaches 45 or 50 years is it able to produce a high quality wine cork. Besides being a sustainable, renewable resource, cork has many special qualities; it is lightweight, relatively impermeable, and elastic. Most cork is used for the wine industry, though it is also used for flooring and insulation.

While, in Portugal, the largest producer of cork, cork forests have been protected since 1259, the increasing popularity of synthetic wine caps are threatening the cork industry and at the same time, the unique cork forests.

_____ ⋞๑∾ _____

Fiber: Textiles and Cloth

Fibers extracted from plants and animals are used to produce textiles and cloth. Cotton (*Gossypium* sp.) is the single most important textile fiber in the world, and accounts for over 40 percent of total world fiber production according to the U.S. Department of Agriculture (USDA 2007). Jute is the second most important natural fiber, and is produced from plants of the family Malvaceae and the genus *Corchorus* spp. Jute is commonly called burlap or hessian, and is used to make sacks, rugs, curtains, upholstery, carpets, rope, cloth, and paper. The earliest fabric known is linen, created from the flax plant (Linum usitatissimum)—has been used for thousands of years—fabric has been found dating from 8000 BC. Flax is historically significant for its use to produce sails. It was the most important fiber until the eighteenth century when cotton replaced it in importance. Some other fibers from plants include hemp (*Cannabis sativa*), sisal (*Agave sisalana*), and ramie (*Boehmeria nivea*). Bamboo can also be processed to create a soft cotton-like fiber. Bamboo fabric has natural antibacterial qualities; unlike cotton it also wicks moisture away from the body like some synthetics. Manufacturers have even developed a fabric from the polymers in corn sugars. A number of fibers are also produced from wood cellulose, including rayon, acetate, and lyocell. Though processing can involve toxic chemicals, some of these fibers can be processed sustainably.

Silk fibers are created from the cocoons of the larvae of silk-worm moths; to make 1 yard of silk cloth requires 3,000 cocoons (see Figure 3.2). There are several species of silkworm moths. The domesticated *Bombyx mori* (Mulberry silkworm) is the most common. Its primary food is leaves of the Mulberry tree (family Moraceae), particularly *Morus alba* (White mulberry). There are other less common varieties, commonly known as "wild silk." For example, *Antherea assama westwood*, endemic to the Brahmaputra valley of India, produces the fine muga silk renowned for its golden color. Silk is well-known for its unique luster and luxurious feel. It is also one of the strongest fibers—the same diameter silk fiber is stronger than a comparable fiber of steel. The unique qualities of silk are difficult to replicate.

Like silk and linen, wool is a minor part of the textile industry. Wool is made from various animal coats; by far the most common comes from sheep. Many other animals produce fine quality, specialty wools

Figure 3.2 Silk fibers are created from the cocoons of the larvae of silkworm moths; to make one yard of silk cloth requires 3000 cocoons (*Horning©CBC-AMNH*. Used by permission)

including angora and cashmere goats, llama, alpaca, and antelopes. Most wool is used to produce clothing though it is also used for carpeting and felt.

Polyester, olefin, nylon, acrylic, and other man-made or synthetic fibers increasingly dominate the textile industry, making up an estimated 60 percent of total worldwide fiber production. Although these fabrics are perceived as not being "natural," it should be noted that (like plastics) they are derived from petroleum, which was originally formed by compressing biodiversity (organic matter) over long periods of time.

Medicine

Biodiversity is essential to human health; extracts from plants and to a lesser extent animals, are critical to treat infections, diseases, and

Figure 3.3 Medicinal plants found at the Mercado de las Brujas in La Paz, Bolivia. Some 80 percent of people depend on traditional medicine in the developing world (*Frey © CBC-AMNH*. Used by permission)

other illnesses. Some 80 percent of people in the developing world rely on traditional medicine, the majority of it from plants (see Figure 3.3). Biodiversity plays a central role in Western medicine, 57 percent of the top 150 most prescribed prescription drugs are either extracted from natural sources or are synthesized based on natural compounds. Biodiversity also forms the basis of medical models that allow us to understand human physiology and disease.

Many Western medicines were developed from a plant or an animal source. Over millions of years, plants and animals have evolved unique compounds that enable them to combat pests or immobilize prey. Molds have been used as medicine for at least 3,000 years; Chinese, Indians, Greeks, and Romans all used molds to treat wounds, inflammation, and infections. It was only later that Alexander Fleming, in the 1920s, made the connection that ordinary bread mold, *Pencillium notatum*, could stop bacterial growth. Later another species, *Pencillium chrysogenum*, was discovered that had more powerful antibacterial activity and forms the basis of the antibiotic penicillin and its derivatives used today to treat many infections from pneumonia to tetanus. Microbes continue to be critical to the development of anti-infective agents. Many antibiotics have

Table 3.4 Some Common Medicines Derived from Plants and Animals

Drug	Source	Use
Barbaloin, aloe-emodin	Aloe (*Aloe spp.*)	Antibacterial, skin conditions, purgative
Atropine	Belladonna (*Atopa belladonna*)	Relaxant, sedative
Codeine	Opium poppy (*Papaver somniferum*)	Painkiller
Colchicine	Autumn crocus (*Colchicum autumnale*)	Anticancer agent
Digitoxin	Common foxglove (*Digitalis purpurea*)	Cardiac stimulant
Ephedrine, Pseudoephedrine	Joint fir (Ephedra sinica)	Asthma, emphysema, bronchiodilator, hay fever
L-Dopa	Velvet bean (*Mucuna deeringiana*)	Parkinson's disease
Menthol	Mint (*Menta spcs.*)	Nasal congestion
Morphine	Opium poppy (*Papaver somniferum*)	Painkiller
Quinine	Yellow cinchona (*Cinchona ledgeriana*)	Malaria
Reserpine	Indian snakeroot (*Rauvolfia serpentina*)	Hypertension
Scopolamine	Thornapple (*Datura metel*)	Sedative
Taxol	Pacific Yew (*Taxus brevifolia*)	Anticancer
Vinblastine, vincristine	Rosy periwinkle (Catharanthus roseus)	Leukemia

Source: Compiled by Laverty.

their origins in soil microorganisms—especially those from the phylum Actinobacteria or Actinomycetes. Other examples include aspirin and common acne medicines derived from salicylic acids, first taken from the bark of willow trees (*Salix sp.*). While many of these drugs are now more efficiently synthesized than extracted from material collected in the wild, we still depend on the chemical structures in nature to guide us in developing and synthesizing new drugs (see Table 3.4).

Some drugs are still synthesized in whole or in part from wild sources. For example, Taxol or paclitaxel, a potent drug originally used to fight ovarian and breast cancers and now used to treat a number of other cancers, was first derived from the bark of the Pacific Yew (*Taxus brevifolia*). In fact, to produce 1 kg of Taxol required 6.7 tons of bark or approximately two to three thousand trees. Because removing the bark killed the trees, and the tree is very slow growing, researchers began investigating alternative sources to create the drug. Fortunately, researchers found that the leaves of the European Yew (*Taxus baccata*), a close relative of the Pacific Yew, produce a similar chemical substance that can be used to produce Taxol more sustainably and less expensively. This semisynthetic production of Taxol or paclitaxel remains partially dependent on needles from wild or cultivated sources of the European Yew (*Taxus baccata*) or the Himalayan Yew (*Taxus yunnanensis*), and has now been replaced with plant cell fermentation. This process is much more sustainable and essentially dependent on plant cell cultures.

Another example is the Indian snakeroot plant (*Rauwolfia serpentina*), which has been used for thousands of years by Hindu healers to treat nervous disorders and mental illnesses. Western scientists only came to recognize its potential as a medicine in the 1940s. The chemical compound extracted from the plant (an alkaloid called reserpine) is a key, active ingredient in drugs that treat hypertension, anxiety, and schizophrenia. Commercial synthesis of reserpine is complex and expensive; it can be extracted naturally for half the cost. *Rauwolfia serpentina* was in such high demand that it virtually disappeared in the wild in India and Indonesia in the 1960s. Related species from Africa and the Americas were found that could act as a substitute and the plant is now also cultivated.

While plants are still the primary source for medicine, the aquatic realm is currently leading the next wave of medical discoveries. Marine organisms produce many novel compounds, including some of the most powerful toxins on earth. For example, the mollusks cone shells (*Conus spp.*) use a special harpoon loaded with a potent venom that paralyzes their prey almost instantly. A new painkiller, Ziconotide or Prialt, created from a peptide from the venom of a cone shell (*Conus magnus*), is not only hundreds of times more powerful than morphine, but it also uniquely targets pain receptors without causing numbness. Injected into the spinal cord, it blocks the channels where pain signals normally travel, basically blocking the pain without causing numbness. There are over 500 species of cone shell and each has 50 to 200 active peptides in its venom; each of these peptides holds the promise of providing the basis for a drug that has a highly specific target point with minimal side effects. Unlike morphine and other opiate painkillers it is not addictive. A compound from a marine sea squirt, *Ecteinascidia turbinata*, has been isolated and is in the last stages of clinical trials for treating cancer. Several other compounds with antitumor properties have been isolated from other marine invertebrates.

A variety of animals are also important as medical models that allow researchers to understand human physiology and disease. Many organisms are ideal for the study of certain organs or diseases. The horseshoe crab (*Limulus polyphemus*) has one of the largest and most accessible optic nerves of any animal. This has made it ideal for the study of animal vision and provided new understanding of vision in humans. Blood from the horseshoe crab also provides the most sensitive test for gram-negative bacteria. The sea squirt or tunicate is the only animal other than humans to develop stones in its kidney-like organ. This animal has been used as a test case to develop strategies to prevent kidney stones in humans.

Indirect Use Values: Services

Ecosystems, and the plant and animal species that constitute them, provide a host of services to all living things, including:

- *regulating global processes*, the regulation of atmospheric gases that affect global and local climates and the air we breathe;
- *soil and water conservation*, maintaining the hydrologic cycle and controlling erosion;
- *nutrient cycling*, the flow of nutrients and energy across the planet—for example, waste decomposition and detoxification, soil renewal, nitrogen fixation, and photosynthesis;
- *a genetic library* that provides a source of information to create better agricultural crops or livestock;
- maintenance of plant reproduction through *pollination and seed dispersal*, in those plants we rely on for food, clothing, or shelter;
- *the control of agricultural pests and disease;*
- *a source of inspiration* to solve agricultural, medical, and manufacturing problems; and
- *tourism and recreation* opportunities.

Often the values of ecosystem services are not considered in commercial market analyses, despite their critical importance to human survival. How can we assign a value to the atmospheric regulation of oxygen? In some ways, its value is infinite since without it we could not survive. Also many ecosystem services cannot be replaced or if they can, it is only at considerable cost. In an attempt to estimate the value of ecosystem services, Costanza and others (1997) calculated that the earth provides a minimum of $16 to $54 trillion dollars worth of "services" to humans per year, based on the value of fifteen ecosystem services and two goods in sixteen biomes. Some scientists and social scientists have disputed the quantitative conclusions of this study; among the criticisms of the study is the difficulty of "scaling up," across ecosystems and for different types of services. Nevertheless, the study was an important first attempt to estimate the global economic contribution of ecosystem services.

Many services provided by biodiversity go beyond what is needed for our immediate survival, including the many cultural, spiritual, and aesthetic values people place on nature and natural areas. Some feel that people have an innate connection or kinship with nature. Nature also provides insight and understanding of our role in the world, and has value for education, as well as for scientific research. Furthermore, each species has an ecological value as part of an ecosystem, and species diversity contributes to ecosystem function and resilience that is the ability for an ecosystem to recover after a disturbance. While species diversity is related to ecosystem function and resilience, there is not necessarily a one to one correspondence. In other words, a hypothetical ecosystem with 150 species is not necessarily twice as good at providing ecosystem services than one with seventy-five species. However, regardless of diversity levels the wholesale removal of species from ecosystems is likely to disrupt the ability of an ecosystem to provide these services.

Global Processes: Atmosphere and Climate Regulation

Life on earth plays a critical role in regulating the earth's physical, chemical, and geological properties, from influencing the chemical composition of the atmosphere to modifying climate.

About 3.5 billion years ago, early life forms (principally cyanobacteria) helped to create an oxygenated atmosphere through photosynthesis, taking up carbon dioxide from the atmosphere and releasing oxygen. Over time, these organisms altered the composition of the atmosphere, increasing oxygen levels, and paved the way for organisms that use oxygen as an energy source (aerobic respiration), forming an atmosphere similar to that existing today.

Over the last century, humans have changed the atmosphere's composition by releasing large amounts of carbon dioxide. This excess carbon dioxide, along with other "greenhouse" gases, is heating up our atmosphere and changing the world's climate, leading to "global warming" or "climate change." There has been much debate about how natural processes, such as the cycling of carbon through phytoplankton in the oceans, will respond to these changes. Will phytoplankton productivity increase and thereby absorb the extra carbon from the atmosphere? Studies suggest that natural processes may slow the rate of increase of carbon dioxide in the atmosphere, but it is doubtful that either the earth's oceans or its forests can absorb the entirety of the extra carbon released by human activity.

Carbon cycles on the planet between the land, atmosphere, and oceans through a combination of physical, chemical, geological, and biological processes. One key way biodiversity influences the composition of the earth's atmosphere is through its role in carbon cycling in the oceans, the largest reservoir for carbon on the planet. Phytoplankton (or microscopic marine plants) regulate atmospheric chemistry by transforming carbon dioxide into organic matter during photosynthesis. This carbon-laden organic matter settles either directly or indirectly (after it has been consumed) in the deep ocean, where it stays for centuries, even for thousands of years, acting as the major reservoir for carbon on the planet. In addition, carbon also reaches the deep ocean through another biological process—the formation of calcium carbonate, which is the primary component of seashells. It is also found in two groups of marine microorganisms: coccolithophorids (a phytoplankton) and foraminifera (a single-celled, shelled organism that is abundant in many marine environments). When these organisms die, their shells sink to the bottom or dissolve in the water column. This movement of carbon through the oceans removes excess carbon from the atmosphere and thus regulates the earth's climate.

Global Processes: Land Use Change and Climate Regulation

The energy source that ultimately drives the earth's climate is the sun. The amount of solar radiation absorbed by the earth depends primarily on the characteristics of the surface. Although the link between solar absorption, thermodynamics, and ultimately climate is very complex, newer studies indicate that vegetation cover and seasonal variation in vegetation cover affects climate on both global and local scales. New generations of atmospheric circulation models are increasingly able to incorporate more complex data related to these parameters. Besides regulating the atmosphere's composition, the extent and distribution of different types of vegetation over the globe modifies climate in three main ways:

- affecting the reflectance of sunlight (*radiation balance*);
- regulating the release of water vapor (*evapotranspiration*); and
- changing wind patterns and moisture loss (*surface roughness*).

The amount of solar radiation reflected by a surface is known as its *albedo*; surfaces with low albedo reflect a small amount of sunlight, those with high albedo reflect a large amount. Different types of vegetation have different albedos; forests typically have low albedo, whereas deserts or snow-covered areas have high albedo. Deciduous forests are a good example of the seasonal relationship between vegetation and radiation balance. In the summer, the leaves in deciduous forests absorb solar radiation through photosynthesis; in winter, after their leaves have fallen, deciduous forests tend to reflect more solar radiation. These seasonal changes in vegetation modify climate in complex ways, by changing evapotranspiration rates and albedo.

Vegetation absorbs water from the soil and releases it back into the atmosphere through *evapotranspiration*, which is the major pathway for water to move from the soil to the atmosphere. Water moves from the soil into the roots and up through the vessels, eventually reaching the leaves, where it evaporates. This release of water from vegetation cools the air temperature. In the Amazon, vegetation and climate are tightly coupled; evapotranspiration of plants is believed to contribute an estimated 50 percent of the annual rainfall in the Amazon. Deforestation in the Amazon starts a complex feedback loop, reducing evapotranspiration rates, which leads to decreased rainfall and increased vulnerability to fire.

Deforestation also influences the climate of cloud forests. For instance, Costa Rica's Monteverde Cloud Forest harbors a rich diversity of organisms, many of which are found nowhere else in the world. However, deforestation in lower-lying lands, even regions over 50 kilometers away, is changing the local climate, leaving the "cloud" forest cloudless. As winds pass over deforested lowlands, clouds are lifted higher, often above the mountaintops, reducing cloud forest formation. Removing the clouds from a cloud forest dries the forest, so it can no longer support the same vegetation or provide appropriate habitat for many of the species previously found there. Similar patterns may be occurring in other, less studied montane cloud forests around the world.

The *surface roughness* of an area is determined by its topography and the type of vegetation that grows there; these features alter the flow of winds in the lower atmosphere and in turn influences climate. Lower surface roughness tends to reduce surface moisture and increase evaporation. Farmers apply this knowledge when they plant trees to create windbreaks. Windbreaks increase surface roughness and thereby reduce wind speed, change the microclimate, reduce soil erosion, and modify temperature and humidity. For many field crops, windbreaks increase yields and production efficiency. They also minimize stress on livestock from cold winds.

Soil and Water Conservation

Biodiversity is also important for global soil and water protection. Terrestrial vegetation in forests and other upland habitats maintains water quality and quantity, and controls soil erosion.

In *watersheds* (a land area drained by a river and its tributaries) where vegetation has been removed, flooding prevails in the wet season and drought in the dry season. With no roots left to absorb rain, soil erosion is more intense and rapid, causing a double effect: removing nutrient-rich topsoil and leading to siltation in rivers downstream and ultimately, the ocean.

The island of Haiti provides a classic illustration of the damage caused by deforestation and erosion. Less than 1 percent of Haiti's forests remain, most of it disappearing in the last 50 years. As a result, when it rains devastating floods stream down the mountains, taking what little soil remains with them, ending up in rivers and eventually the sea. The lack of water being absorbed prevents the island's aquifers from being replenished. The flooding has destroyed two-thirds of the island's farmland since 1940—the combined effect of lost topsoil, and irrigation canals filling up with sediment. The environmental crisis has become a health crisis with near-starvation and disease.

Siltation also damages coastal environments. Soil is loaded with extra nutrients; while this sounds like it might be a good thing, it is the extra nutrients entering the water that leads to excess algae growth. The extra productivity reduces oxygen, especially as the algae dies and creates massive "dead zones"—areas in which there is no oxygen and nothing can live. Coastal areas with low oxygen (hypoxic) or no oxygen (anoxic) are an issue throughout much of the United States and the world (Long Island Sound, New York; Penascola Bay, Florida; Chesapeake Bay, Virginia; Baltic Sea, Northern Europe; Black Sea, northern Adriatic). The Black Sea was once the largest "dead zone" in the world caused by agricultural runoff, the area recovered as fertilizers became too costly once the Soviet Union collapsed. Currently the largest "dead zone" is in the Gulf of Mexico, where the Mississippi River enters; it covers 20,000 square kilometers (or 7,722 square miles).

One of the most productive ecosystems on earth, *wetlands* have water present at or near the surface of the soil or within the root zone all year or for a period of time during the year, and the vegetation there is adapted to these conditions. Wetlands are instrumental for the maintenance of clean water and erosion control. Microbes and plants in wetlands absorb nutrients that would otherwise go directly into coastal waters, and in the process filter and purify water of pollutants before they enter other aquatic systems. Wetlands also reduce flood, wave, and wind damage. They retard the flow of floodwaters and accumulate sediments that would otherwise be carried downstream or into coastal areas. Wetlands also serve as breeding grounds and nurseries for fish and support thousands of bird and other animal species.

Nutrient Cycling

Nutrient cycling is yet another critical service provided by biodiversity. Fungi and other microorganisms in soil help break down dead plants and animals, eventually converting this organic matter into nutrients that enrich the soil.

Nitrogen is essential for plant growth, and an insufficient quantity of it limits plant production in both natural and agricultural ecosystems. While nitrogen is abundant in the atmosphere, only a few organisms (commonly known as nitrogen-fixing bacteria) can use it in this form. Nitrogen-fixing bacteria extract nitrogen from the air and transform it into ammonia, then other bacteria further break down this ammonia into nitrogenous compounds that can be absorbed and used by most plants. In addition to their role in decomposition and hence nutrient cycling, microorganisms also detoxify waste, changing waste products into forms less harmful to the environment.

A Genetic Library for Crops and Livestock

Humans cultivate only a small fraction of the plant and animal species on earth (see Figure 3.4). We depend on biodiversity as a genetic library to sustain existing agricultural systems. Using genetic material from wild counterparts of crops or domesticated animals, we can create new breeds that are more tolerant of pests and disease, or that are more suited to certain environments.

In the late 1970s, teosinte (*Zea diploperennis*), the closest wild relative of corn, was discovered and found to be resistant to viral diseases that infect domesticated corn, *Z. mays*. Teosinte has the same chromosome number as *Z. mays* and can therefore hybridize with it. When this occurs, some of the viral resistance is transferred to domestic corn. Four viral-resistant commercial strains have since been produced, highlighting the importance of wild counterparts to cultivated food crops.

There are other cases in history when a widely grown crop has failed due to disease, with devastating consequences. One famous example is the Irish potato famine, which led to the death of a million people. In the mid-nineteenth century, a blight (or fungus-like pathogen) destroyed

Figure 3.4 Many crops are now grown as vast monocultures, with one species extending over large areas, such as this rice field in Vietnam (*Frey © CBC-AMNH*. Used by permission)

much of the crop. European potato crops were particularly susceptible to infection since they had originated from only a few sources and thus were genetically very similar. To combat this disease, a long search began to find a plant resistant to the blight. By the early twentieth century, a related plant in Mexico provided the solution. Hybridizing this plant with potatoes produced a resistant strain. Unfortunately, this was not a permanent solution. Today, potato blight is once again a concern, and it is likely the solution lies in existing biodiversity. As the world's crops become increasingly homogenized, it is important to remember the lessons of this event: agricultural systems with higher genetic diversity are often more resilient, and ultimately, biodiversity may solve these crises.

Pollination

About 75 percent of flowering plants depend on pollinators such as bees, wasps, various insects, birds, and bats, to reproduce (see Figure 3.5). While some plant-pollinator relationships are specialized, many plants are visited by a diverse array of pollinators. Each gain from this mutualistic relationship—the plant is fertilized, while the pollinator gets pollen, nectar, or a similar reward. Plants and their pollinators are increasingly threatened around the world. This has been well observed

Figure 3.5 About 75 percent of flowering plants depend on pollinators such as bees, wasps, various insects, birds, and bats, to reproduce (*Ersts©CBC-AMNH*. Used by permission)

in Europe where pollinators have been monitored for many years, but similar trends are emerging in North America. Pollination is critical to most major agricultural crops and virtually impossible to replace. For instance, imagine how costly fruit would be (and how little would be available) if natural pollinators no longer existed and each developing flower had to be fertilized by hand.

DISAPPEARING BEES

About a third of U.S. crops are pollinated by western honeybees (*Apis mellifera* L.). Beekeepers typically rent out their bees to farmers to pollinate their crops. Honeybees are popular as managed pollinators as they are generalists and forage over large distances. Beginning in the fall of 2006, beekeepers noticed something amiss with the world's most popular pollinators—billions of bees were becoming disoriented, unable to return to their hives. Termed "colony collapse disorder," the phenomenon claimed around a quarter of bee colonies in thirty-five U.S. states between the fall of 2006 and 2007, with similar occurrences showing up worldwide. Scientists suspect the disease may be caused by a virus, fungus, or mite, and hope to unravel the mystery soon before crops are significantly impacted. While theories abound, it is difficult to separate out the cause from the symptoms.

Seed Dispersal

> Nature, like a careful gardener, thus takes her seeds from a bed of a particular nature, and drops them in another equally fitted for them. (Charles Darwin, *On the Origin of Species*, 1866, p. 462)

Many animal species are important dispersers of plant seeds. While scientists hypothesize that the loss of a specific seed disperser could cause a plant to become extinct, to date there is no definitive example where this has occurred. A famous example often cited is the case of the dodo (*Raphus cucullatus*) and the tambalacoque tree (*Sideroxylon grandiflorum*). The dodo, a large flightless bird that inhabited the island of Mauritius in the Indian Ocean, became extinct due to overhunting in the late seventeenth century. It was once thought that the germination of the hard-cased seeds of the tambalacoque tree depended upon the seeds passing through a dodo's gizzard. By the 1970s, only thirteen trees remained and it was thought the tree had not reproduced for 300 years. As an experiment, the trees hard-coated seeds were fed to a turkey; after passing through its gizzard the seeds became viable and germinated. Further research showed that several other species (including three now-extinct species—a large-billed parrot, a giant tortoise, and a giant lizard) besides the dodo were capable of cracking the seed coat. Thus, it wasn't just the dodo but the loss of several species that led to the decline of the tambalacoque tree.

Unfortunately, the decline and eventual extinction of species are often unobserved and so it is difficult to tease out the cause of the end result, as multiple factors often operate simultaneously. Similar problems exist today in understanding current population declines. For example, in a given species, population declines may be caused by loss of habitat, loss in prey species, or loss of predators, a combination of these factors, or possibly some other yet unidentified cause, such as disease.

In the pine forests of western North America, corvids (including jays, magpies, and crows), squirrels, and bears play a role in seed dispersal. The Clark's nutcracker (*Nucifraga columbiana*) is particularly well-adapted to dispersal of whitebark pine (*Pinus albicaulis*) seeds. The nutcracker removes the wingless seeds from the cones, which otherwise would not open on their own. Nutcrackers hide the seeds in clumps. When the uneaten seeds eventually grow, they are clustered, accounting for the typical distribution pattern of whitebark pine in the forest.

In tropical areas, large mammals and frugivorous birds play a key role in dispersing the seeds of trees and maintaining tree diversity over large areas. For example, three-wattled bellbirds (*Procnias tricarunculata*) are important dispersers of tree seeds of members of the Laurel family (Lauraceae) in Costa Rica. Because bellbirds return again and again to one or more favorite perches, they take the fruit and its seeds away from the parent tree, spreading Laurel trees throughout the forest.

Natural Pest Control

Agricultural pests (principally insects, plant pathogens, and weeds) destroy an estimated 37 percent of U.S. crops (Pimentel et al. 1995). Destruction varies depending on the crop, where it is grown, and the type of pest. According to a study by Oerke and others (1994), for example, average production losses for the principal cereals and potatoes due to pests, pathogens, and weeds amount to 15 percent, 14 percent, and 13 percent respectively. Without natural predators that keep pests in control, these figures would be much higher. Pesticides are no replacement for the services provided by these crop-friendly predators. Pesticides reduce water quality and impact farmer and food safety. Pests develop resistance, which leads to a constant need to apply more pesticide or find new pesticides to battle pests. Pesticides are expensive; natural pest control saves farmers billions of dollars each year. They not only kill harmful pests but other insects that are beneficial (such as natural predators or pollinators).

Farmers have looked for more sustainable alternatives. Tools like integrated pest management where farmers monitor pest populations and aim to reduce their reliance on pesticides is one approach. Natural pest control is another alterative. Lady bugs (Coleoptera: Coccinellidae) are one of the best-known natural predators. They consume aphids, scale, mealy bugs, white flies, and spider mites. They are popular for biological control agents particularly to control aphids on crops.

Ecological Value

Natural communities are dynamic systems in which component species have complex interrelationships. Species that have ecological roles that are greater than one would expect based on their abundance are called *keystone species*. Removal of one or several keystone species may immediately have ecosystem-wide consequences, often continuing decades or even centuries later. Ecosystems are complex and difficult to study, thus it is often difficult to predict *a priori* which species are keystone species.

Species that are important due to their sheer numbers are often called *dominant species*. These species make up the most biomass of an ecosystem. For example, salt marshes that extend along the Atlantic and

Gulf coasts of North America are dominated by several grass species, including *Spartina patens* (saltmeadow cordgrass) and *Spartina alterniflora* (smooth cordgrass) among other species.

The ability of a community to adapt and respond to changing environmental conditions is complex and has created much scientific debate. How many species does an ecosystem need to function? Which ones are essential and which can be removed and still maintain productivity? In theory, species diversity should increase an ecosystem's stability (ability to withstand a disturbance) and resilience (ability to recover from a disturbance). In practice, sometimes this is the case, and at other times this direct relationship cannot be demonstrated. This may be partially explained by the fact that many ecosystems have built-in redundancies so that two or more species' functions overlap. Because of these redundancies, some changes in the number or type of species may have little impact on an ecosystem. On the other hand, if enough species are removed, there is a good chance of disrupting ecosystem function. In addition, it is extremely difficult to predict which species are redundant and what the effect of removing any one species would be on a system.

PACIFC KELP FORESTS

Kelp forests, as their name suggests, are dominated by kelp, a brown seaweed of the family Laminariales. They are found in shallow, rocky marine habitats from temperate to subarctic regions, and are important ecosystems for many commercially valuable fish and invertebrates.

Vast forests of kelp and other marine plants existed in the northern Pacific Ocean prior to the eighteenth century. The kelp was eaten by herbivores such as sea urchins (Family Strongylocentrotidae), which in turn were preyed upon by predators such as sea otters (*Enhydra lutris*). Hunting during the eighteenth and nineteenth centuries brought sea otters to the brink of extinction. In the absence of sea otters, sea urchin populations burgeoned and grazed down the kelp forests, at the extreme creating "urchin barrens," where the kelp was completely eradicated. Other species dependent on kelp (such as red abalone *Haliotis rufescens*) were affected too. Legal protection of sea otters in the twentieth century under the International Fur Seal Treaty (1911) led to a partial recovery of the thirteen remnant sea otter populations.

More recently, sea otter populations in western Alaska have plummeted. Among the possible reasons for the decline include starvation from lack of prey, predation by killer whales (*Orcinus orca*), migration, disease, and pollution. Populations of Stellar sea lions (Arctocephalus townsendi), Harbor seals (*Phoca vitulina*), and northern fur seals (*Callorhinus ursinus*) in the area are also declining. It remains to be seen what is causing the decline in sea otter and other marine mammal populations in this region.

Interestingly, Southern California's kelp forests did not show immediate effects after the disappearance of sea otters from the ecosystem. This is thought to be because the system was more diverse initially and thus perhaps more resilient. Other predators (California sheephead fish, *Semicossyphus pulcher,* and spiny lobsters, *Panulirus interruptus*) and competitors (abalone *Haliotis spp*) of the sea urchin may have helped maintain the system. However, when these predators and competitors were over-harvested as well in the 1950s, the kelp forests declined drastically as sea urchin populations boomed.

In the 1970s and 1980s, a sea urchin fishery developed that then enabled the kelp forest to recover. However, it left a system with little diversity. The interrelationships among these species and the changes that reverberate through systems as species are removed are mirrored in other ecosystems on the planet, both aquatic and terrestrial.

As this example illustrates, biodiversity is incredibly complex and conservation efforts cannot focus on just one species or even on events of the recent past.

Biodiversity As a Model for Technological Advancement

We value biodiversity for its ability to inspire creativity and to help us solve problems. The word *Biomimicry* was coined to describe how the natural world has inspired people to solve problems in agriculture, medicine, manufacturing, and commerce. Humans have long drawn inspiration from the wild:

- *Velcro* was patterned after cockleburs, a plant that has spiny seeds that attach to clothes as people walk through a meadow.

- A closer look at hedgehog spines, whose supple, strong structure enables them to bend without breaking, led to the development of *lightweight wheels* in which the tires have been replaced with an array of spines that effectively absorb shock.

- Millipedes, invertebrates with multiple pairs of legs fringing their long bodies, are being studied to help design *robots* to carry heavy weights in cramped conditions where significant twisting and turning is necessary.

- Scientists study primates such as baboons, chimpanzees, and howler monkeys in the wild to learn how they "self-medicate" against diseases and how they use compounds from plants to regulate processes such as reproduction. This information can help scientists in the search for new drugs for humans.

- Studies of the structure and function of natural grasslands are revealing new methods for fertilizing crops and protecting them from pests.

- Biodiversity also serves as a model for medical research that allows researchers to understand human physiology and disease. When bears hibernate during the winter, they stop most normal functions (such as eating, drinking, urinating, or defecating) for 150 days. But unlike some hibernating animals, bears only lower their temperature slightly to accomplish this feat. Researchers are trying to understand the physiological changes that allow them to survive. One discovery is a blood protein that slows organ metabolism and reduces blood coagulation. This protein could be given to trauma patients being rushed to the hospital after a severe accident to minimize blood loss. Bears also recycle urea when they hibernate. In humans a toxic buildup of urine is fatal in a matter of days but bears seem able to break down urea and reuse it to build tissue. Most animals that don't exercise—including people—lose bone, but bears survive hibernation with little to no bone loss; rather than losing calcium they can extract and continue to maintain bone mass. Understanding these mechanisms could lead to treatments for kidney disease or osteoporosis.

Scientific and Educational Value

Biodiversity provides a way for us to understand the world. Darwin's observations of the diversity of finches on the Galapagos led to the development of his theories of evolution and natural selection. Curiosity about the world around us is fundamental not only to scientific investigations, but also to education (see Figure 3.6). The diverse species and

Figure 3.6 Biodiversity provides a way for us to understand the world. This image shows a sampling quadrat being laid by divers on a coral reef in the Bahamas (*Brumbaugh©CBC-AMNH*)

ecosystems around us provide unique educational opportunities and models to explore and learn from.

THE BACTERIA THAT REVOLUTIONIZED GENETIC RESEARCH

When scientists first explored the hot springs of Yellowstone National Park in the 1960s, they thought life couldn't survive above 55°C/131°F. They were surprised to discover that the springs were indeed "alive" with bacteria. Among the bacteria they found was *Thermus aquaticus*, which thrives at temperatures of 70°C or 160°F. *Thermus aquaticus* is a member of the ancient group of bacteria—the Archaea—that were here when the Earth formed, and when the environment on the planet was similar to that of the hot springs. The bacteria's enzymes can also withstand extremely high temperatures. One of these enzymes, taq polymerase, has become the mainstay of genetic research. Taq polymerase catalyzes the polymerase chain reaction (PCR), which allows you to make billions of copies of DNA in just a few hours. PCR has revolutionized genetic research, opening new possibilities for improved health, agriculture, and even criminology (see Figure 3.7).

Figure 3.7 Mudpots and geysers in Yellowstone National Park. It was during explorations of the hotsprings there that scientists discovered life could survive at temperatures greater than above 55°C/131°F (*National Park Service (NPS). Photo by William S. Keller, 1970*).

Tourism and Recreation

Natural areas such as forests, lakes, mountains, and beaches provide venues for commercially valuable outdoor activities such as ecotourism, bird watching, sport fishing, hunting, and hiking. Economists have estimated that the total recreational value of the world's resources could be as high as $800 billion annually. The growing ecotourism industry generates an enormous amount of money and is fast becoming a lucrative industry for some developing nations. For example, in Costa Rica, tourism has expanded rapidly since the mid-1980s and is the leading source of foreign revenue, surpassing the banana industry in 1999, reaching $1.92 billion USD in 2007.

Cultural and Spiritual Value

Plants and animals are central to mythology, art, literature, dance, song, poetry, rituals, festivals, and holidays around the world. Even governments identify with certain animals, such as the eagle, an emblem for Egypt, Germany, Mexico, Nigeria, and the United States, and the bear, Russia and Finland's national animal. Sports teams often use images of birds, insects, and mammals as their official mascots. Advertising campaigns incorporate animals into numerous commercials in the hopes of attracting our attention to sell their products. The natural world also influences areas as diverse as housing styles, type of dress, and regional farming methods.

The relationship between people and nature is reflected in language. Many metaphors use plants and animals, such as "happy as a clam" or "busy as a beaver." In many Asian cultures, bamboo has a central role in arts and traditions and is common in proverbs, for example, "Make sure your life is as pure and straight as a bamboo flute." Plants and animals represent different things in different cultures. Rats are considered pests in much of Europe and North America, delicacies in many Asian countries, and sacred in some parts of India. Of course, within cultures individual attitudes also vary dramatically. For example, in Britain most people don't like rodents; yet there are many hobbyists who devote their life to breeding them, including members of the National Fancy Rat Club and the National Mouse Club.

Edward O. Wilson coined the term, *Biophilia*, to describe the concept that a love of nature was ingrained in our genes. He argues that our very survival through human evolution depended on a detailed knowledge of natural history. This was vital for finding food, shelter, materials, and medicine. Those lacking the ability to develop an intimate

Figure 3.8 The natural world has contributed to the development of human spiritual traditions (*Frey©CBC-AMNH*. Used by permission)

knowledge of biodiversity were simply less successful at surviving and reproducing.

The natural world has also contributed to the development of human spiritual traditions (see Figure 3.8). Religions help define the relationships between humans and their environment. Many religious guidelines arose as a response to natural phenomena, particularly threatening ones. Rules to protect crops, sustain fisheries, or avoid danger developed from the natural world, punishment was meted out through crop failures or other threats to survival; the reward was a healthy life and continued survival.

The idea of kinship between animals and people is common in many parts of the world. It is an important concept in animist belief systems; where all objects are believed to have souls or spirits. In some places, animist beliefs extend to how natural resources are managed. Certain acts are prohibited leading to punishment or a series of taboos. These taboos often apply to vulnerable natural resources. One example comes from the Lowland Peruvian Amazon; local fishers avoid fishing in particular oxbow lakes and seasonal fishing areas, because of a belief that there are spirits that protect all fish, animals, and plants found there. The most common belief is that these areas have "mothers" (usually giant anacondas (*Eunectes murinus*), caiman (*Caiman crocodiles*), jaguars (*Panthera onca*), or various species of monkey) who will kill anyone who tries to extract resources. Research has shown that these lakes are important breeding areas for the fish. Another example is from the islands of the South Pacific, fishing families have a unique relationship with certain animals, usually turtles or sharks. For each family, these special species (or groups of species) are considered sacred and it is taboo to hunt them; this relationship is carried on through the generations.

Animals feature in myths around the world. The tortoise in North America bears the world on its back—its upper shell represents heaven, its lower shell the underworld, and its middle the earth. A similar tortoise image appears in the myths of China. In North America, Thunderbirds represent the upper world of spirits, while serpents live beneath the earth. The Lightning Bird and the Rainbow serpent are found in the myths of Central and Southern Africa.

Nature is used in religious imagery, and many religious traditions view the contemplation of nature as an important spiritual value:

- In Thailand, trees are marked with yellow cloth to denote their sacredness to the Buddhist faith. This practice has saved some sacred groves from illegal logging, since to destroy these trees is a severe crime.
- In Japan, Shinto temples are often located in large groves of trees, where spiritual forces are believed to exist.
- Human stewardship over plants and animals (God's other creations less able to protect themselves) is a central tenet of Christian, Judaism, and Islam. For example, the Garden of Eden and Noah's Ark are important symbols that define human responsibility for biodiversity.

Aesthetic Value

Gazing at a stunning mountain landscape, a tropical beach, a beautiful sunset, or a majestic lion, nature often evokes feelings of awe (see Figure 3.9). In studies conducted in developed countries, when people are shown two landscapes, and asked to pick the one they prefer, people invariably selected natural scenes as more beautiful than the urban ones. People also value species for their beauty, rarity, complexity, and variability.

People are frequently attracted to natural areas for recreation or relaxation, or as a source of inspiration. It is common for people to decorate their homes and office with plants. Many hobbies reflect people's fascination with nature, such as birdwatching, hiking, gardening, scuba

Figure 3.9 People also appreciate nature for its aesthetic beauty (*Ersts©CBC-AMNH*. Used by permission)

diving, or even watching nature shows. Zoos and botanical gardens offer a unique opportunity for people to view rare species that they would otherwise only be able to see in books or on television. In 2000, 134 million people visited zoos and aquariums in the United States, more than attended all professional baseball, football and basketball games combined, according to the American Zoo and Aquarium Association (AZA). A study conducted by the AZA of over 5,500 visitors, showed that people who visit zoos and aquariums feel a stronger connection to nature and are more motivated to become involved in conservation solutions (Falk et al. 2007). Their visits reinforced existing beliefs of stewardship of nature and a love of animals.

Non-Use Values

Non-use or passive values are used to categorize the value that people place on biodiversity that do not have a direct use. The two principle values usually included in non-use values are: *existence value*, the value of knowing something exists even if you will never use or see it, and *bequest value*, the value of leaving something behind for the next generation. *Potential or option value* is sometimes included in this category, and refers to the value of something that has not yet been recognized, such as the potential or future value of a plant as a medicine. Some authors consider this a form of use value, as it focuses on the option to use something in the future; alternatively this value can be considered a *non-use value*, as it is uncertain what or if something will have a potential use. Bequest value, similarly, is sometimes considered a use value, as it is leaving behind something that the next generation can use. Another value that is often overlooked is the *strategic value* of a species for conservation and scientific research. Panda bears, tigers, and whales are all common symbols used in the conservation movement.

Because of their abstract nature, measuring these values is complex and thus it is more difficult to incorporate them into conservation decisions. Economists sometimes use surveys to estimate these values, for example, asking how much people are willing to pay to save the tiger for future generations (bequest value) or just to know that the tiger exists (existence value). There are many economic-based techniques employed to evaluate biodiversity, such as cost-benefit analyses, safe minimum standard, willingness to pay, willingness to accept, and contingent value.

Potential Value

An important consideration in weighing the importance of biodiversity is to acknowledge that there is much we do not yet understand. Who would have guessed that the bark of the Pacific yew would become an important cancer-fighting agent? Who can guess what will be valuable to us in the future? This argues for retention of as much biodiversity as is possible on the theory that it may have a potential value for something we have not yet recognized. Aldo Leopold, an American naturalist and the "father" of the field of wildlife ecology, is one of the most famous supporters of the idea that wildlife and wildlands are valuable in and of themselves. Leopold (1887–1948) noted:

> If the biota, in the course of aeons, has built something we like but do not understand, then who but a fool would discard seemingly useless parts? To keep every cog and wheel is the first precaution of intelligent tinkering.

Leopold professes a precautionary approach that suggests there are many things that may have a potential value in the future, which we have yet to recognize. The precautionary approach has become a guiding principle for much environmental planning and legislation. A telling example of the importance of biodiversity as an insurance policy for an unforeseen difficulty is provided by the International Rice Research Institute (IRRI). The IRRI keeps a seed bank of grain varieties that are no longer commercially produced, and as a result was able to forestall the spread of a disease that negatively affected rice yields.

———————————————— ✑ ————————————————

SEEDBANKS—SAVING THE WORLD'S PLANT DIVERSITY

Rice production may be the most important economic activity on the planet. Rice farms cover 11 percent of the world's arable land and provide daily food to over half the world's population. In some countries, it represents 60 to 75 percent of all calories consumed. What would happen if production decreased significantly?

In the early 1970s a disease—the Grassy Stunt Virus—emerged along with its carrier, the brown plant hopper, and infested rice crops in much of Asia. The newly developed, high-yield varieties of rice that were in use were not resistant to the virus and decimated harvests.

The International Rice Research Institute (IRRI), created in 1960, was able to help because of their program to preserve the genetic diversity of rice plants. The IRRI maintains a seed bank of about 100,000 types of rice, including recently developed hybrids and wild varieties no longer in production.

Over 6,000 different varieties were tested for genetic resistance to the Grassy Stunt Virus before a resistant variety (*Oryza nivara*) was discovered. Using crossbreeding, a new strain of rice (IR36) was developed that had high productivity and was resistant to the virus. The new strain was introduced in 1976 and is now grown around the world. This highlights the role of conserving biodiversity as insurance for agriculture.

———————————————— ✑ ————————————————

Existence and Bequest Values

Many people will never see a panda bear or a tiger, yet they are happy to know that these species exist. The value people place on just knowing a species or ecosystem exists, even if they will never see it or use it, is known as *existence value*. There are many possible motivations behind this desire. Economists use surveys to put a monetary estimate on this value, by asking questions such as how much people are willing to pay to save a particular species from extinction. This measure is controversial as often these questions are hypothetical. Bequest value, or the value of knowing that a species or ecosystem will be there for your children or grandchildren to see or use, is similarly difficult to measure.

Strategic Value

Genes, species and ecosystems also have strategic value to conservation biologists. As noted above, conservation biologists strategically use charismatic species, such as the tiger, as a way to increase support for conserving biodiversity in general. Scientists also use species strategically as indicators of ecosystem health. Indicator species allow scientists to measure and monitor an ecosystem's health efficiently. For example, certain aquatic invertebrates are used to monitor for water pollution.

Intrinsic Value

> Every form of life is unique, warranting respect regardless of its worth to man. (United Nations Charter for Nature, 1982)

Intrinsic value is the inherent worth of something, independent of its value to anyone or anything else. One way to think about intrinsic value is to view it as similar to an inalienable right to exist. The Endangered Species Act in the United States protects many species that are not "valuable" to humans in any readily definable way (for instance, the dwarf wedge mussel (*Alasmidonta heterodon*) or the swamp pink flower (*Helonias bullata*)). These species are protected based on the idea that they have a right to exist, just as all humans do.

Intrinsic value is a frequently misunderstood term as some consider values that are not easily defined, such as aesthetic values, to be intrinsic values. However, as discussed earlier, aesthetic values are better considered as a kind of extrinsic value, because aesthetic values provide humans with a service of sorts—our own satisfaction. Others consider a species' value to the structure and function of an ecosystem (such as an invertebrate decomposer's ability to cycle nutrients) as its intrinsic value because it does not have any obvious value to humans. However, this ecosystem value is still utilitarian value, except it focuses on one organism's usefulness to another organism, rather than to humans.

The concept of intrinsic value is highly philosophical. Many economists and some ethicists believe that intrinsic value does not exist, arguing that all values are human-centered, and that a value cannot exist without an evaluator.

Generally, two contrasting beliefs frame a continuum along which our beliefs fall:

- On one end is the idea that humans are the center of the universe and that nature exists (and is used) for human benefit (a view called *anthropocentrism*).
- At the other end is the notion that life is the center of the universe and humans are a separate but equal part of nature (*biocentrism, or ecocentrism*).

The biocentric view, forwarded by the deep ecology movement, holds that all species have intrinsic value and that humans are no more important than other species. Thus everything has an equal right to exist simply because it already exists. Having this right will result in also having a "right" to have one's future survival guaranteed to an extent equal to any and all other species. If one accepts the idea that biodiversity has intrinsic value, then species conservation requires less justification. In other words, if a species is intrinsically valuable, regardless of its use to humans or to other species, it should be conserved, and then the onus is on those who do not want to conserve the species to provide a justification for its removal. Intrinsic value is a central tenet of many religions. For example, many of the world's largest religions, including Christianity, Judaism, Islam, Buddhism, and Hinduism, consider everything on earth to be inherently sacred, or sacred as a result of being created by a divine being, and thus, intrinsically valuable, and humans are responsible to care for and respect these creations.

Why Do Values Matter?

Ultimately, every decision people make, consciously or not, is based on what they, as individuals, value. As Mark Sagoff (1988) writes, "if individuals in the future have no exposure to anything we consider natural or unspoiled, they will not acquire a taste for such things. What they want will be more or less what we leave to them."

Values are central to conservation decisions, and conservation biology has even been termed a "value-laden" science. When we measure biodiversity or set conservation priorities, we must decide which species, populations, or ecosystems to study, monitor, manage, or conserve, and these choices depend upon what we currently value.

- Which species/ecosystems should be protected? Should we give priority to a species/ecosystem that is nationally endangered but globally common, or to a species/ecosystem that is nationally common and globally rare?
- Should we value areas with greater numbers of species over those with many endemic species (those that are found only in one place in the world)?

There are no correct answers to these questions—the responses depend upon what people value and what information is available to make these decisions. Values are also the basis of arguments used to justify the conservation of species or ecosystems, for example is a particular area valuable for recreation, logging, or fishing.

In most countries, conservation efforts focus on species listed as endangered and threatened, although to date, these lists include mainly vertebrates and vascular or higher plants. Since we know so little about other components of biodiversity (invertebrates, nonvascular plants, microbes, etc.), our current endangered species lists may be omitting information critical to better decision-making about our imperiled species. Also, people are often biased toward "charismatic" species, such as lions or panda bears. A poll in the United States (Czech and Krausman 1997) determined that the public values plants, birds, and mammals more than all other groups (fish, reptiles, amphibians, invertebrates, and microorganisms). They also found that the public considered ecological importance and rarity as the key reasons to conserve species. A study by the National Center for Ecological Analysis and Synthesis in the United States in 2002 raised the question of whether there might also be an inadvertent scientific bias toward "cute, unique, or spectacular" species. Indeed endangered species recovery plans for vertebrates got significantly more attention than those for invertebrates and plants.

4

THE STATE OF THE
WORLD'S BIODIVERSITY

This chapter discusses the state of the world's biodiversity and links this to some of the major threats to biodiversity. It covers not only the direct threats to biodiversity, such as the disappearance of habitats, but also some of the underlying factors that contribute to the loss of biodiversity, such as weak legislation and lack of enforcement.

INTRODUCTION

Humans dominate the planet to an extent never before seen. Our rapidly expanding populations and economies place staggering demands on the world's limited resources. To meet these needs, about one-third of the planet's land surface has been substantially altered by human activity. Many species barely manage to survive on a fraction of their former range and in increasingly fragmented landscapes. Grasslands and tropical dry forest ecosystems have almost completely disappeared from the planet, replaced by farmland. Dams disrupt freshwater ecosystems, while overfishing, pollution, and habitat destruction threatens the marine world. Our planet is increasingly made up of species that can only survive in human-modified landscapes. Humans are also transporting plants and animals around the globe, both deliberately and unintentionally. These "invaders" threaten other species or change entire ecosystems. Human influence reaches the farthest corners of the globe: the Arctic and Antarctic are contaminated by pollutants created tens of thousands of kilometers away and carried through the air. Human industrial activity has even altered the earth's atmosphere, through the release of carbon dioxide, and these changes have begun to change the earth's climate.

Only by understanding the threats to biodiversity can we hope to conserve it. The leading direct threats to biodiversity include ecosystem

Table 4.1 Threatened and Endangered Species

	Number of Described Species	Number of Species Evaluated By IUCN	Number Threatened in 2007 At % of Total Described Species	Number Threatened As % of Species Evaluated By IUCN
Mammals	5,416	4,863	20%	22%
Birds	9,956	9,956	12%	12%
Reptiles	8,240	1,385	5%	30%
Amphibians	6,199	5,915	29%	31%
Fishes	30,000	3,119	4%	39%
Total Vertebrates	59,811	25,238	10%	23%
Invertebrates	1,203,375	4,116	0.18%	51%
Plants	297,326	12,043	3%	70%

Note: Note few invertebrates are evaluated for risk of endangerment compared to the total described species.
Source: IUCN Red List 2007.

fragmentation, invasive species, pollution, overexploitation, and global climate change. The underlying causes of biodiversity loss, on the other hand, are often more complex and stem from many interrelated factors, the most important are overpopulation and unsustainable consumption of resources. Existing socioeconomic structures and policies contribute to biodiversity loss and hinder conservation efforts by reducing incentives to conserve. Corruption, weak governance and legislation, and a lack of enforcement further exacerbate the threats to biodiversity.

According to the IUCN Red List of Threatened Species, scientists have assessed conservation status for fewer than 10 percent of known species. Of the vertebrate species assessed about 23 percent are considered threatened. Examining individual vertebrate groups, 12 percent of birds are at risk, 22 percent of mammals, 30 percent of reptiles, 31 percent of amphibians, and 39 percent of fish as of 2007 (see Table 4.1). Those species that rely on freshwater habitats are typically the most threatened.

───── ∽∞∾ ─────

DISAPPEARING CYCADS

There are about 300 species of cycads, many of them discovered in the last 25 years. Cycads resemble palms or ferns, but are in their own division: the Cycadophyta. Found in tropical and subtropical regions around the world, they are an ancient plant dating as far back as the Permian, 200 million years ago. Though they were common in the Jurassic Period, there are only about 305 species of cycad remaining today. These ancient origins make them particularly interesting to scientists.

Twenty-five percent of cycads are endangered or critically endangered. Cycad populations are at risk because of their habitat is disappearing, and collectors seek them out as they are an attractive plant. An additional problem for their survival is their biology; cycads are long lived and slow growing, so populations take a long time to establish.

———————————————— ∽∾ ————————————————

Ecosystem Loss and Fragmentation

Today, "the fragmented landscape is becoming one of the most ubiquitous features of the tropical world—and indeed, of the entire planet." (Laurence and Bierregaard 1997, p. xii). Many scientists consider the disappearance and fragmentation of the world's ecosystems the greatest worldwide threat to wildlife and the primary causes of species extinction (see Figure 4.1).

Ecosystem loss and fragmentation are related processes and typically occur at the same time. While fragmentation is sometimes defined as the loss and isolation of natural habitats, scientists prefer to view the two processes—loss and fragmentation—as distinct from each other. In Chapter 1, we learned that an *ecosystem* is comprised of an assemblage of organisms and the physical environment in which they exchange energy and matter. *Ecosystem loss* refers to the disappearance of an entire ecosystem—that is the total loss of all components of the system—that is the landscape as well as the species that live there. A clear example would be cutting down a forest and replacing it with agriculture; here the forest ecosystem has been lost entirely.

Many researchers examine loss with respect to a specific organism's *habitat*, and refer to "habitat loss and fragmentation" rather than "ecosystem loss and fragmentation." Habitat loss and fragmentation is related to the loss of a specific animal's habitat whereas ecosystem loss and fragmentation is all-encompassing, referring to the loss of all the different species found there.

There are two common definitions of *habitat*. The first defines habitat as a species' use of the environment, while the second as an attribute of the land and refers more broadly to a *habitat* that supports a diverse assemblage of species. In this book, we use *habitat* and *habitat type* to differentiate between the two common usages of the word *habitat*. *Habitat loss* is the modification of an organism's environment to the extent that the qualities of the environment no longer support its survival. Habitat loss usually begins as *habitat degradation*. In other words, before a species'

(a)

(b)

Figure 4.1 Loss and fragmentation of ecosystems is a concern around the world. (a) (above) Agriculture fragments tropical forest in southern Bolivia's Amazon region habitat (Santa Cruz, Bolivia) (Sterling©CBC-AMNH). (b) Agriculture fragments temperate forest in New York's Catskill region (*Frey©CBC-AMNH*. Used by permission)

habitat disappears completely usually the quality of its habitat declines. Once the habitat's quality has become so low that it no longer supports that species, then it is termed *habitat loss*. A species might depend on large trees in a forest. If these are selectively removed the forest might still remain but the habitat has disappeared for species that depend on large trees. In this situation, the habitat type (the forest) might be relatively intact, but the habitat for the species that depends on particular trees might have disappeared entirely.

Fragmentation is usually a product of ecosystem loss and is best thought of as the subdivision of a formerly continuous landscape into smaller *patches* or pieces of land. For example, as a forested landscape becomes fragmented, the landscape becomes a series of small forest patches surrounded by a *matrix* of habitat often unsuitable to wildlife (such as agricultural land). Ultimately, fragmentation isolates populations by preventing species from moving across the landscape or migrating, and this in turn leads to genetically isolated populations.

Loss and fragmentation impact most of the earth's major *biomes*. A biome is a major biotic classification characterized by similar vegetation structure and climate, but not necessarily the same species. The leading land classification was developed by Robert G. Bailey, initially for the United States in 1975, and later the world in 1989. The classification is centered on four major domains: Dry, Humid Tropical, Humid Temperate, and Polar. These are further subdivided: Mediterranean Forests and Shrubland, Deserts and Shrubland, Montane Grasslands and Shrublands, Temperate Grasslands and Savanna, Tropical Grasslands and Savanna, Tropical and Subtropical Coniferous Forests, Temperate Coniferous Forests, Tropical Broadleaf Dry Deciduous Forest, Tropical Broadleaf Humid Deciduous Forest, Temperate Broadleaf Deciduous Forest, Boreal Forest or Taiga, and Tundra. In different parts of the world biomes often have local names, so the Temperate Grasslands of the western United States are known as Prairies, while in Argentina, they are called Pampas, and in Asia, they are called Steppes.

Biomes that are especially vulnerable to loss and fragmentation are typically those in greatest demand for human use. Grasslands, for example, exhibit some of the highest rates of loss and fragmentation due to their suitability for growing wheat and corn, and for grazing livestock (see Figure 4.2). An estimated 20 to 50 percent of nine of the world's 14 major biomes have been converted to crops according to the Millennium Ecosystem Assessment (see Figure 4.3). About 35 percent of coastal mangrove forests have been lost around the world, much of that in the last decade with the rise of shrimp farming (see Figure 4.4).

Figure 4.2 Grasslands have some of the highest rates of loss and fragmentation of any ecosystem, as the land is also suitable for wheat, corn, and raising livestock (*Photo by Jeff Vanuga, USDA Natural Resources Conservation Service*)

Mangroves are specially adapted to live in salt water. In many places they are a key component of tropical coastal systems, where they protect the coast from storms and provide nursery habitat for young fish. The Tundra and Boreal Forest, not surprisingly, still remain relatively untouched by people.

Many of the world's major rivers are highly fragmented; people have interfered with their flow largely to redirect water for different uses. According to the World Register of Dams, between 1950 and 1986 the number of large dams (either higher than 15 meters (49.2 feet) or with a volume exceeding 3 million cubic meters (or 3.9 million cubic yards) built to meet water and energy needs, topped 45,000. While dams do provide irrigation water and generate power, they also fragment habitat, disrupt fish migration, and alter natural water flow patterns, often with irreversible impacts. In the Pacific Northwest of the United States, salmon (*Salmo* spp.) populations declined dramatically during the nineteenth century, a result of fishing pressure combined with mining and dam construction, and their populations remain in a precarious state. Dams seriously limit the recovery of salmon populations by preventing

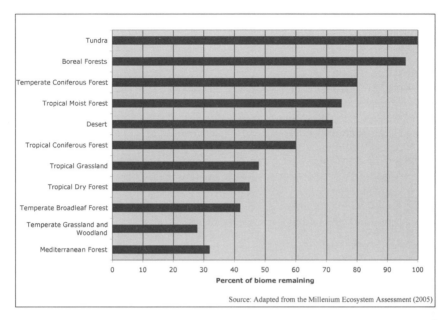

Figure 4.3 Between 20 and 50 percent of nine of the world's 14 major biomes have already been converted to crops (Adapted from *Millennium Ecosystem Assessment*)

Figure 4.4 Mangroves are disappearing around the world. In many places they are a key component of tropical coastal systems, where they protect the coast from storms and provide nursery habitat for young fish (*Ersts©CBC-AMNH.* Used by permission)

Table 4.2 Some Examples of Dams Recently Removed Or Scheduled to Be Removed in the United States

- Milltown Dam, Milltown, Montana [2008]
- Marmot Dam, Sandy River, Oregon [2007]
- Horse Creek Dam, Horse Creek, California [2006]
- South Batavia Dam, Fox River, Illinois [2005]
- Marvel Slab Dam, Cahaba River, Alabama [2004]
- Sturgeon River Dam, Sturgeon River, Michigan [2003]
- Sennebec Dam, St. George River, Maine [2002]
- Franklin Dam, Sheboygan River, Wisconsin [2001]
- Rockdale Dam, Koshkonong Creek, Wisconsin [2000]
- Edwards Dam, Kennebec River, Maine [1999]

Dams Scheduled to be Removed
- Glines Canyon Dam, Elwha River, Washington
- Condit Dam, White Salmon River, Washington
- Clarkston area, Snake River, Washington
- Klamath River, California
- San Clemente, California
- Veazie Dam and Great Works Dam, Penobscot River, Maine
- Wittlinger Dam, Yellow Breeches Creek, Pennsylvania
- Dillsboro Dam, Tuckasegee River, North Carolina
- Woodside I and II dams, Twelvemile Creek, South Carolina

Source: The number of dams removed each year was: 20 (1999), 28 (2000), 23 (2001), 44 (2002), 34 (2003), 37 (2004), 34 (2005), 33 (2006), 54 (2007). [Figures based on *American Rivers. Restoring Rivers,* See http://www. americanrivers.org/site/PageServer?pagename=AR7_RestoringRivers_Pubs.]

them from returning to their natal streams to reproduce and altering water quality. You might wonder why you can still find salmon in the supermarket. Alaskan populations appear healthy, and about 80 percent of North American wild salmon is harvested there. Salmon is also increasingly farmed in coastal areas.

Large dam construction had its heyday in the 1950s when major dams were built in the United States and many developed nations. While large dams are still being constructed in some parts of the world, recent trends in the United States have favored removing dams instead. About 750 dams have been removed since 1912 according to the American Rivers Organization. The rate of removal is accelerating, as many dams licenses in the United States are expiring, and not being renewed by the Federal Energy Regulatory Commission (see Table 4.2).

Large dams are still being constructed in many parts of the world; the river basins with the most dams being built are the Yangtze (China), the Tigris and Euphrates (Iraq, Syria, Turkey), and the Danube (shared by 19 countries including Austria, Croatia, Germany, Hungary, Romania,

Slovakia). Once completed in 2009, China's Three Gorges Dam on the Yangtze River will be the world's largest, stretching 1.3 miles across and reaching 600 feet in height. More than one million people will be relocated to make way for the dam and reservoir. The Three Gorges Dam is expected to generate 18,000 megawatts of power. In comparison, the world's second largest dam, Itaipu, located on the border between Brazil and Paraguay produces 12,600 megawatts of power, while the third largest dam, the Grand Coulee in Washington State, generates just 6,480 megawatts of power. Hydroelectric dams supply energy that is relatively clean compared to coal-generated power. However, they do have drawbacks, as they permanently alter the environment where they are put in place, and their ability to generate power over the long term depends on the nature of the river. Over time the reservoirs begin to fill with sediment, which limits the power of the dams.

After a large dam is built, huge areas are flooded turning a river into a lake or reservoir. When the Tucuruí dam was constructed on the lower Tocantins River in northeastern Brazil, 2,850 km^2 of tropical rainforest was flooded. Following the creation of the dam, decaying vegetation in the reservoir as well as the reservoir's shape severely diminished the quality of the water flowing downstream. As oxygen levels in the water declined, many species could no longer survive. In addition, the flooding led to a rise in mosquito populations, which then appeared to increase the rate of malaria transmission in the surrounding communities (World Commission on Dams, 2000).

Large dams can also fail leading to major environmental and human catastrophes. This is becoming an increasing risk as dams start to age and are not repaired. Recent examples of major breaches and failures include: the Zeysoun Dam break in Syria, which released 20 million tons of water killing 20 people and leaving thousands homeless (2002); the Taum Sauk Dam in Missouri, United States, released 1 billion gallons of water in just 12 seconds (2005); the Shakidor Dam in Pakistan breached after excessive rain, killing at least 80 people (2005); Hawaii's Ka Loko Dam produced 18-foot waves after a 250-meter break in the dam and killed seven people (2006); and flooding breached a dam along the River Danube in Romania that endangered thousands of people (2006).

Effects of Fragmentation

Fragmentation and loss of ecosystems are linked processes; fragmentation is a consequence of loss. It is difficult to distinguish the effects of one process from the other as they often occur simultaneously. Ecosystems can disappear without causing fragmentation if all the loss occurs in one area; however, more often than not when an ecosystem is removed usually it causes some fragmentation. For example, if a forest is completely removed for a suburban development it may not cause fragmentation if there is no surrounding forest; the result may simply be ecosystem loss. Alternatively, if forest is removed to put in a road but the forest remains alongside the road, then you have both ecosystem loss and fragmentation.

Loss of ecosystems impact species principally by reducing available resources and microenvironments. Fragmentation has additional consequences for species on top of those caused by loss, the most important being that it impacts species' movement and dispersal as well as their behavior. As a landscape becomes fragmented, three major consequences are apparent: decreasing size of the patches of remaining forest or other ecosystems; increased edge effects; and increased isolation of patches.

Decreasing Patch Size

Once a landscape is fragmented, it leaves behind "patches" or "islands" of ecosystems. The size of the remaining patches is a critical factor in determining the number and type of species that can survive within them. For all species that cannot or will not leave a patch, all requirements to complete their life cycle must be met within it, from finding food to choosing mates. This is especially important for species with complex life cycles and that have distinct habitat requirements. For example, many frogs, salamanders, and toads breed in ephemeral ponds or wetlands in the spring and depend on them as tadpoles, but later as adults they live in the surrounding woodlands. Both the wetlands and woodlands are essential for their survival, and need to be found within a patch for them to survive.

Large patches typically support larger populations of a given species and thus reduce the risks of: *extinction, inbreeding depression* (the reduction in reproductive ability and survival rates as a result of breeding among closely related individuals), and *genetic drift* (a random change in allele frequency in small breeding populations leading to a loss of genetic variation). Some species, such as the scarlet tanager, require large areas of continuous habitat and cannot survive in small patches; they are referred to as *area-sensitive species*.

Increased Edge Effects

Once a landscape has been fragmented you are left with patches of habitat that are now bordering another habitat, such as farmland or a road, the area where these two habitats meet are known as *edge environments.* Many studies have examined the *effects of edges* on the physical environment and biological communities that remain after fragmentation. The longest running and perhaps the most detailed study of fragmentation effects ever conducted is the Biological Dynamics of Forest Fragments project, which began in 1979. This pioneering project, located in the Amazon region north of Manaus, Brazil, has informed much of our general understanding of the effects of forest fragmentation.

"Edge effects" is a general term used to describe changes along the boundary between two contrasting habitats. These impacts may be physical; changes in wind and sunlight along an edge influence what trees and plants survive there. The creation of edge environments also directly influences the species' composition, abundance, and distribution, and indirectly changes species' interactions such as predation, competition, pollination, and seed dispersal. Moreover, many of the effects of fragmentation are synergistic, in other words their combined impact is greater than their individual impact; for example, fragmentation can lead to increased fire risk, increased vulnerability to invasive species, or increased hunting pressure.

Increased Patch Isolation—Barriers to Dispersal

The degree of isolation of a patch helps determine what biological communities it can sustain. While patches may appear isolated, their actual biological connectivity, or the degree to which patches in a landscape are linked, depends on the dispersal capability of the species found there. For some species, crossing an open field for 2 kilometers is not a problem. However, species that spend most of their time in treetops (e.g., some species of primates and marsupials) or in dark, interior forest may never cross such a large opening. A species that disperses over long distances, such as an African elephant (*Loxodonta* sp.), will perceive a particular landscape as more connected than a species with short-range dispersal, such as a shrew (species of the family Soricidae). In a very isolated patch, species that cannot disperse may be unable to find adequate resources or mates. They may become separated from other populations and are thus prone to genetic inbreeding and possibly local extinction.

FRAGMENTATION AND ECOLOGICAL PROCESSES: SIMPLIFIED SYSTEMS

After fragmentation some species disappear and are replaced by more common species; biologists call this process *faunal relaxation.* Large vertebrates, especially those at higher *trophic levels*, are particularly susceptible to loss and fragmentation, and among the first species to disappear. A trophic level is a stage in a food chain or web. Primary producers, such as plants, are at the lowest trophic level, through to herbivores (species that eat only plants), and primary and secondary carnivores (the highest trophic level). Thus, predators often disappear from a landscape before their prey, and those species that do survive on small fragments (usually herbivores) become far more abundant than populations of the same species on larger species-fragments. An example is the disappearance of wolves and an increase in the number of deer.

There are two principal explanations for the increased abundance of herbivores. The first is *ecological release from competition*: when competing species are removed, the resources they used become available to the surviving species. The second is that prey can now escape predators that normally limited their abundance on larger fragments. This lack of predators in small fragments can lead to an overabundance of herbivores that tend to "weed out" plant species and convert the landscape into a forest of "herbivore-proof" plants.

Furthermore, as large predators disappear, smaller predators often increase; this is known as *mesopredator release*. In California, the disappearance of coyotes from fragments resulted in an overabundance of smaller predators, such as skunks, raccoons, gray fox, and cats. These smaller predators then prey on scrub-breeding birds. Fragmentation thus triggers distortions in ecological interactions that drive a process of species loss, the end point of which is a greatly simplified ecological system lacking much of the initial diversity.

Species Vulnerable to Fragmentation

In a given landscape, the effects of connectivity and isolation vary greatly from species to species. Species that fly (i.e., birds, bats, flying insects) and naturally occur at higher densities are much less affected by patch isolation than less mobile species that tend to rely on less available habitats and naturally occur at lower densities (i.e., frogs and beetles).

A species' response to fragmentation depends on its perception of the environment. Species that can't fly or climb to get an aerial view of a landscape make decisions primarily based on the habitat they see directly in front of them. A study conducted in the Amazon revealed distinct responses of similar animals to fragmentation. Two species of opposum— the wooly (*Didelphys lanigera*) and the mouse (*Didelphys murina*)—were tracked using radio transmitters to determine if they would travel a gap of 135 to 275 meters to reach the fragment of forest on the other side. Mouse opossums were able and willing to cross the gap, while the wooly opossum, which spends most of its time in the trees, was not.

Behavior, resource needs, reproductive biology, and natural history are indicators of species that are most vulnerable to fragmentation

(Laurence and Bierregaard, 1997). Organisms exhibiting one or more of the following characteristics are often highly affected by fragmentation of their environment:

- rare species with narrow distributions and/or already small populations;
- species with large home ranges, such as top carnivores or large animals;
- species that need heterogeneous landscapes with diverse habitats;
- species that avoid matrix habitats (the area between fragments such as open farmland) or that have very specialized habitat requirements;
- species with limited dispersal abilities or low fecundity; and
- coevolved species (i.e., plants that need specific pollinators).

Natural versus Human Fragmentation

Landscapes are fragmented naturally and by people. There are a number of key differences between these two causes of fragmentation.

1. A naturally patchy landscape often has a *complex* structure with many different types of patches. A human-fragmented landscape tends to have a *simplified* patch structure with more distinct edges, often with a few small patches of natural habitats in a large area of developed land.
2. Patch types in human-modified landscapes are often *unsuitable* to many species, while in a heterogeneous natural landscape most patch types are *suitable* to a more diverse group of species.
3. The borders (or edges) of patches in naturally patchy landscapes tend to be *less abrupt* than in those created by humans. There is a transition between different environments.
4. Certain features of human-fragmented landscapes, such as roads, are *novel* in the evolutionary history of most wild species and pose unusual threats. In addition to the direct threat that heavy traffic presents, roads fragment the environment and make remote areas more accessible to hunters and traders. They also facilitate the introduction of nonnative and potentially invasive species.

INVASIVE SPECIES

Invasive species are the second greatest threat to biodiversity conservation globally, threatening individual species as well as entire ecosystems. Moreover, as humans carry species from one part of the world to the other, we potentially interfere with evolutionary processes—the ultimate source of the world's biodiversity. Spatial barriers are a fundamental driving force of evolution and by transferring species repeatedly across natural species barriers we disrupt the evolutionary process. The frequency, geographic scope, and sheer number of plants and animals carried by people from one area to another has increased tremendously as transportation and commerce have progressed.

―――――――――――― ∽∂∾ ――――――――――――

GALÁPAGOS ISLANDS

The Galápagos Islands of Ecuador is famous as a laboratory of evolution, driven primarily by the isolation of the chain's 127 islands from the mainland and from one another (see Figure 4.5). One hundred and fifteen terrestrial vertebrates, 560 plants, and an estimated 2,000 invertebrates are native to the islands. While the islands were untouched by humans until their discovery in 1535, there are now 24 species of introduced vertebrates, over 700 introduced plants, and an unknown number of introduced invertebrates. These latest additions can also thank the dramatic increase in tourists to the islands. Beginning with just one ship of 66 visitors in 1967, tourism to the Galapagos has expanded rapidly. Even in recent years the growth has been dramatic, increasing from 69,000 to 150,000 visitors between 2000 and 2006. To meet this tourist demand, the local population has surpassed 20,000 people. Each year, five cargo ships bring 33,000 tons of materials to the islands, while over 100 vessels move between five ports, 50 visitor

Figure 4.5 The Galápagos Islands of Ecuador, famous as a laboratory of evolution, are threatened by growing tourism (*Cullman*)

sites, and uncounted fishing areas, covering nearly 3 million nautical miles within the archipelago. Obviously, for an archipelago that witnessed almost no humans before 1700, this loss of isolation has major consequences for the native biota. The Galápagos Islands was the first site added to UNESCO's World Heritage list in 1978; however, in 2007 it was listed as in danger due to threats from expanding tourism, increasing human population, invasive species, and lack of institutional support to combat these threats.

— ∽⌢∾ —

Multiple terms are used to describe invasive species. While many of these terms are synonymous and can be used interchangeably, it should be noted that there is a major distinction between *exotic* species and *invasive* species. An *exotic species* lives outside its native range. Terms such as "nonindigenous," "nonnative," "alien," "adventive," "neophytes" (for plants only) and "introduced" are synonymous with "exotic." *Invasive species*, on the other hand, can be either *exotic* or *native* species, whose populations have expanded dramatically and thus out-compete, displace, or extirpate native species, potentially threatening the structure and function of intact ecosystems. Not all exotic species become invasive. Many populations of exotic species do not survive for long in their new environment. For example, tropical species rarely survive a temperate winter so they will not become established. Others become established but do not substantially disrupt their new host environment. Similarly, not all invasive species are exotic. Scientists are increasingly documenting native species whose populations grow out of control or substantially increase their range due to changes in their environment. These range shifts may be due to natural or human-induced change as it can be difficult to distinguish between the two. These species' populations expand as they often prey on or parasitize species at a higher rate than before. Alternatively, these native invasives may out-compete close relatives, or mate with them to create hybrids.

Less is known about *native invasives* than exotic ones, but some examples from North America include the brown-headed cowbird (*Molothrus ater*) and the coyote (*Canis latrans*) in the eastern United States. Cowbirds are native to North America but were once restricted to the Great Plains, where they followed bison herds, eating insects turned up by bison hooves. After bison disappeared, these highly adaptable birds began following livestock instead, and expanded their range as humans spread across the landscape. They can now be found throughout the United States, subarctic Canada, and northern Mexico. Cowbirds impact other native birds, as they are brood parasites; they lay their eggs in the nests of other bird species rather than build their own nests. These birds then raise the cowbird young, often to the detriment of their own offspring.

The coyote is native to the western plains of North America. Coyotes are generalists and thrive even in suburban areas. Considered an invasive in the eastern United States, coyotes were first observed in the region in the 1940s and 1950s, where they have largely replaced the niche once occupied by the wolf. Eastern coyotes are often larger than their western counterparts, sometimes feeding on deer. Some suspect that Eastern coyotes are in fact hybrids with wolves, which are now extinct from the eastern United States. In the opposite direction, Saltmarsh or Smoot cordgrass (*Spartina alterniflora*), native to saltwater wetlands of the East coast of North America, was introduced to San Francisco Bay in the 1970s and has altered the entire ecosystem. It is now found along the entire Pacific coast of the United States.

Research into why and how some species become invasive is still preliminary in part because of the complex process of invasion. Predictive models still in development detail the species most likely to become invasive and the potential consequences of their invasion. Research suggests there are three major stages in the process of invasion by exotic species: dispersal, establishment, and integration.

INVASION PROCESS

Dispersal Stage

The first stage—the dispersal stage—comprises how species move from one area to another. The relentless globalization of human societies has afforded exotic species myriad avenues for dispersal into new environments. Humans have unwittingly brought "stowaways" along in containers such as ship hulls and ballast water, on muddy shoes, or in our digestive tracts that have profoundly changed the face of their new environment. Classic examples of these stowaways include:

- *rodents*, such as the Norway rat (*Rattus norvegicus*) and black rat (*Rattus rattus*), and the house mouse (*Mus musculus*)—that have contributed to the demise of innumerable native species around the world, most notably island species unaccustomed to such competition;
- *diseases* (such as small pox and measles) that decimated indigenous human populations when carried to the New World and Australia by European explorers and colonists;
- *marine invaders*, such as European green crabs (*Carcinus maenas*) that may have arrived in the in ship bilge water, and that are changing the structure of inter-tidal communities along the west coast and in freshwaters of the Great lakes;

- and *plant invaders*, such as the southern Russian or Ukrainian leafy spurge (*Euphorbia esula*) that arrived in the United States as a contaminant of grain and is now crowding out the remaining native plant species in prairie grasslands.

Humans deliberately brought exotic species with them when they settled in new lands. These species served as sources of food, commerce, fiber, fuel, medicine, sport, scientific interest, windbreaks, and for enjoyment. Sometimes species have been introduced to help contain outbreaks of other invasive species, unfortunately often with unanticipated consequences.

Domestic cats are potent introduced predators in many parts of the world. In the United States, there are 58 million domestic cats, estimated to kill over 200 million birds each year. Cats are partly responsible for the endangerment of at least six species of North American birds and small mammals, and for the extinction of more than 20 animal species in Australia. An infamous example of a destructive cat was a pregnant cat belonging to a lone lighthouse keeper on Stephen's Island, located between the North and South Islands of New Zealand. Cats quickly overran the island and, evidently hunted and killed every last individual of the Stephen's Island wren (*Traversia lyalli*), a nocturnal, flightless bird, ironically just as it was identified as a species new to science.

How successful an exotic species is at colonizing an area depends on where the species originated, the characteristics of the region colonized, and how the species is transported to the new location.

Establishment

The second stage in the process of invasion—establishment—refers to how biological and physical factors in the colonized region affect the *initial* survival, reproduction, and expansion of invasive species in a new area. An overwhelming majority of exotic species is unsuccessful in establishing populations when introduced to mainland settings. Some scientists argue that tropical oceanic islands are more susceptible to invasions, but the evidence is mixed. Many organisms arrive in a new region and are swiftly eliminated by other species, disease, or simply the climate. Others survive, but do not expand their populations dramatically or become harmful to the environment.

Integration

The final stage in the invasion process—integration—embraces how exotic species interact with the communities and ecosystems that they invade, and the factors that affect their rates of expansion, that is, increases in their population size and the area where they live. As previously noted, some exotic species settle into their new environment and become permanent residents. They do not depend upon re-immigration from their natural range to persist. A few of these species then become invaders. Scientists estimate that, of every 1,000 species that reach a new region, around 100 will settle temporarily, ten will establish long-term populations, and one will become a problematic invasive species. The transition from naturalized immigrant to invader often encompasses a long delay, or "lag phase," followed by a phase of exponential population growth that diminishes only when a species reaches the boundaries of its new range. Many immigrant populations disappear during this time. It is difficult to predict which species will remain as naturalized immigrants and which will become invasive.

WHAT MAKES AN INVASIVE SPECIES SUCCESSFUL?

Unfortunately, it is practically impossible to predict the success of a particular invasive species; nor can we come up with a comprehensive list of attributes of common invaders. Some evidence points to greater success for species with higher numbers of initial invaders, for those that are widespread in their native habitat, as well as for those that settle in already heavily disturbed areas. The European starling (*Sturnus vulgaris*) was released by Eugene Scheiffelin in New York's Central Park. Sixty birds were released in 1890 and another 40 birds in 1891. Some maintain this was part of an effort to bring all the birds mentioned in the writings of Shakespeare to North America. From these relatively modest beginnings, the European starling has become arguably the most pervasive invasive bird in the United States. Fifty years after its release, its population size was estimated at 120 million birds. Starlings are displacing native birds, particularly cavity nesters throughout the eastern United States.

In terrestrial ecosystems, the most successful intruders seem to be those that are significantly different from native species. For example, *Myrica faya*, a tree brought by the Portuguese from the Azores islands, has successfully established itself in Hawaii and is disturbing entire ecosystems. The exotic tree, a member of the legume family, harbors symbiotic bacteria that convert atmospheric nitrogen to ammonia, a trait not present in native Hawaiian plants. These nitrogen-fixing capabilities enable the tree to colonize nutrient-poor volcanic sites faster than native plants. The introduced tree then alters soil chemistry, preventing native species adapted to nitrogen-poor soils from establishing. It also forms dense canopies beneath which other plants don't grow.

ECOLOGICAL CONSEQUENCES OF INVASIVE SPECIES

The list of consequences of invasive species on their host environment is as lengthy as it is depressing. Invasive species can cause local or global extinctions as well as complete disruption of an ecosystem's structure and function. As predators, invasive species often benefit from encountering "naïve" prey that have not yet developed appropriate defenses. A well-known example of this is the brown tree snake (*Boiga irregularis*), which was introduced inadvertently on many Pacific islands and subsequently caused the extinction of a number of native birds, bats, and lizards. Invasive herbivores, such as cows, goats, pigs, and rabbits, also benefit from plants that have not adapted to them. Similarly, newness of exotic species to a region may also mean that they themselves escape predation, as potential predators do not yet look upon them as prey. Freed from their natural predators, competitors, and diseases, populations of exotic species flourish in their new environment.

Exotic invasive species often out-compete native species for food, water, shelter, nutrients, light, and space. The North American gray squirrel (*Sciurus carolinensis*) is out-competing and replacing the native red squirrel (*Sciurus Vulgaris*) in Britain and mainland Europe. Particularly successful invaders include zebra and quagga mussels (*Dreissena polymorpha* and *Dreissena bugensis*), which were introduced into the Great Lakes region of North America sometime in the 1980s. These mussels can achieve densities of up to 627,000 per square yard (or 750,000 per square meter) or higher, blanketing whole lake bottoms and other surfaces. In Lake Erie, where zebra mussels extirpated a healthy population of native, freshwater bivalves (Unionoidea), some shells were covered by 15,000 zebra mussels—the equivalent of five times the weight of the living bivalve. Zebra mussels are extremely efficient filter feeders and, as a result, substantially modify the aquatic systems they invade. They have had a tremendous economic impact as well, blocking water intake pipes as well as coating boats and navigational buoys.

Invasive species also impact native species abundance through hybridization. In North America, Mallard ducks (*Anas platyrhynchos*) have spread through introduction by sport hunters into new areas and by expansion of their range as natural areas are converted to agricultural lands. As mallards encounter closely related black ducks (*Anas rubripes*) in these new areas, they interbreed with them, often coming to dominate the gene pool of smaller populations, as is also the case with the Mexican duck and the mottled duck. North American Mallards are also hybridizing with the New Zealand Gray Duck and the Hawaiian Duck (*Anas wyvilliana*).

Though not often considered in discussions of invasives, diseases are a special class of invasive species that affect wildlife and human populations worldwide. In the Hawaiian Islands, avian pox and malaria have led to an almost complete extermination of endemic birds in lowland forests. Similarly, emerging infectious diseases may play a role in global amphibian decline. Amphibian declines may be in part due to Chytrid fungi that grow in the animals' skin and thereby suffocate them, as well as other viruses and bacteria. Microorganisms are easily transported to new areas, for example in waters used to transport aquarium fish for the pet trade, or even on the nets and boots used by herpetologists themselves.

Invasive species transform whole ecosystems by affecting fertility, productivity, and stability. For instance, plant invaders alter ecological processes such as fire regimes, nutrient cycling, and hydrological cycles or they directly replace the dominant species in a community. In the fynbos ecosystem of South Africa's Cape Province, native plant species have evolved to withstand the harsh dry environment. These plants process water very efficiently so they are able to sustain themselves through the dry summer months. They can survive in nutrient-poor soils and their roots bind the soil, thus minimizing erosion. Finally, much of the vegetation have fire resistant leaves or bark, and their overall low biomass minimizes the impacts of fires that occasionally move through the area. In contrast, eucalyptus, pine, acacia, and other invasive plants in the fynbos scrubland are heavy water users. They have increased the overall biomass and water demands in the ecosystem and raised fire intensity. This has not only threatened the extinction of many endemic plants, but also reduced the amount of water for agricultural production and for the cities of Cape Town and Port Elizabeth.

Invasive species seem to be particularly successful in establishing themselves in and significantly changing the structure and function of freshwater lakes and stream ecosystems. Introduction of game or commercial fish in lakes and streams around the world has wreaked havoc on local fish species. The San Francisco Bay and Delta was rated the most invaded aquatic system in North America by the U.S. Fish and Wildlife Service. In many bay and coastal communities, from Canada to Mexico, exotic species outnumber native species.

CONTROLLING INVASIVE SPECIES

Biotic invasions can have devastating economic consequences. First, they affect potential economic output, causing losses in crop and livestock production, and fisheries profits. Second, the cost of battling invasions (including invasives that are threats to human health)—from quarantine to control to eradication efforts—is enormous. Accurate assessments of these costs are difficult to calculate, but estimates exceed $138 billion USD per year.

There are two major ways to limit invasive species. The first is to prevent new invasions, and the second is to minimize their impact once they have colonized. Initial invasions can be minimized by using quarantine. Quarantine techniques are hampered by our inability to predict which species might become invasive. In addition, countries such as the United States and Australia apply an "innocent until proven guilty" approach to incoming species, mostly to avoid limits on trade. In other words, all species are let in until we know they are harmful. A major problem with this approach is that once we know a species is harmful, it is often already too late to control its spread.

Attempts to control invasive species have focused on chemical (e.g., herbicides), mechanical (e.g., hand removal of giant land snails), and biological (e.g., introduction of native parasites) methods. Each of these has a suite of problems associated with it ranging from cost (mechanical) to side effects of applications (chemical and biological). Control efforts have been most successful at the beginning of an invasion, when populations are small and localized. Once populations take hold and produce significant numbers of offspring, any attempt at control is usually unrealistic, financially and logistically. Frequently, invasive species control efforts are stymied by public concern either for the species being eradicated or the proposed method. For example, some invasive species such as mute swans or feral cats, elicit tremendous public sympathy even as they exact dramatic harm on ecosystems. Considerable effort should be directed at educating the public about invasive species and their effects on biodiversity.

UNSUSTAINABLE EXPLOITATION

The resources humans consume to survive and attain various levels of comfort exert tremendous pressure on the world's plants and animals. While direct use of wildlife and other natural resources is essential for human survival, their overexploitation is a critical problem in conservation. Though habitat loss may be the greatest threat to most species, the overexploitation or nonsustainable use of wildlife is closely linked and plays a powerful role in the loss of biodiversity. Over-harvesting, nonsustainable use, and illegal trade of some species is threatening not only their continued survival, but also that of ecosystems, and the livelihoods of communities and local economics that depend upon these species.

A History of Overexploitation

There is no question that overexploitation has led to species extinctions in historic as well as modern times. Unsustainable hunting led to the near disappearance of the bison. A similar fate has plagued many species of whales, while logging or gathering of wild populations has also brought many plants close to the brink. Biologists use different terms to describe the extent a species may be considered "extinct." *Global extinction* signifies that no living individuals of the species remain anywhere in the world, while *local extinction* refers to the disappearance of a species from a particular area. *Commercial extinction* occurs when populations are too depleted or scattered to be harvested economically, while *ecological extinction* indicates populations that may still be present in low numbers, but no longer play important functional roles in the ecosystem. Overexploitation is one of the suspected causes for the historical global extinctions of the elephant birds and giant lemurs of Madagascar, the giant kangaroos of Australia, the moa of New Zealand and the megaherbivores of North and South America. More recently, overexploitation has resulted in the ecological extinction of large animals, such as jaguars, woolly monkeys, and curassows, from many parts of neotropical forests of Central and South America; this has been termed the "Empty Forest" syndrome; the forest remains but devoid of animals.

According to IUCN 784 species have gone extinct since 1500, 18 of these species since the year 2000. The Pyrenean ibex was once found in large areas of Spain north to France's Pyrenees. Eventually restricted to Ordesa National Park in northern Spain, the last ibex died there in 2000 (see Table 4.3). The year 2000 also marked the first documented primate extinction since 1800; the ultimate cause of the extinction of Miss Waldron's red colobus monkey (*Procolobus badius waldroni*) was attributed to hunting. The monkey was originally restricted to Ghana and the Ivory Coast. Since the declaration a few hunters have brought forward remains of what appear to be colobus monkeys, so it is possible they still exist in very small numbers. This highlights the difficulty in officially declaring something extinct. The almost certain extinction in the wild of the Alagoas Curassow (*Mitu mitu*) from northeastern Brazil in the late 1980s was caused by the interaction of habitat loss for sugarcane fields and "ceaseless hunting," just as hunting was a significant factor in local extinctions of the green peafowl (*Pavo muticus*) throughout Southeast

Table 4.3 Selected Recent Species Extinctions

Common Name	Scientific Name	Year Declared Extinct	Last Known Location
Western Black Rhinoceros	*Diceros bicornis longipes*	2006	Northern Cameroon
Po'o-uli (a bird)	*Melamprosops phaeosoma*	2004	Maui, Hawaii
Miss Waldron's Red Colobus Monkey	*Procolobus badius waldronae*	2000–2001	Ghana
Pyrenean Ibex	*Capra pyrenaica pyrenaica*	2000	Northern Spain
Golden Toad	*Bufo periglenes*	1989	Costa Rica
Arcuate Pearly Mussel	*Epioblasma flexuosa*	1988	North America
Atitlan Grebe,	*Podilymbus gigas*	1987	Guatemala
Kaua'i 'O'o,	*Moho braccatus*	1987	Kauai, Hawaii
Alagoas Currasow	*Mitu mitu*	1980s	Northeastern Brazil
Eungella Gastric-brooding Frog	*Rheobatrachus vitellinus*	1985	Australia
Aldabra bush-warbler	*Nesillas aldabrana*	1983	Seychelles
Javan Tiger	*Panthera tigris sondaica*	1980	Java, Indonesia
Southern Day Frog	*Taudactylus diurnus*	1979	Australia
Dutch Alcon Blue (butterfly)	*Maculinea alcon arenaria*	1979	Netherlands
Colombian Grebe	*Podiceps andinus*	1977	Colombia
Round Island Burrowing Boa	*Bolyeria multocarinata*	1975	Mauritius
Toolache wallaby	*Macropus greyi*	1972	Australia

Source: Adapted from IUCN, and Birdlife International.

Asia and its extirpation from peninsular Malaysia. Local and ecological extinctions of overexploited species can be expected to remain undetected until they become the focus of investigations, by which time it is often too late. Occasionally species that are believed extinct are rediscovered, since many only remain in remote locations, evidently with very low populations. These are known as "Lazarus species." One of the most famous recent rediscoveries was of the Ivory-billed Woodpecker (*Campephilus principalis*) in Arkansas, though evidence is still disputed since it was first sighted.

In theory some level of exploitation should be manageable. The difficulty is in determining what level is manageable over the long-term and in keeping exploitation at that level or below. As in other aspects of conservation, short-term perspectives often call for higher rates of use than long-term ones. Moreover, the harvest theory is often too simplistic to address the complexities of natural systems. Maximum sustained yield theory, the most commonly used approach, assumes that after harvest a population will increase its reproductive rate to compensate for lost individuals, but this is often not the case. Natural variability in populations and environmental oscillations also mask the effects of overexploitation. Depletion is often blamed on "natural cycles" when overexploitation might be to blame. Furthermore, industry and governments (via subsidies) rapidly invest in harvesting (and therefore increase exploitation) during periods of abundance and are reluctant to divest (to achieve a lower exploitation rate) during periods of scarcity. Often government and politicians have a short-term outlook and therefore feel the need to support an unsustainable industry rather than a long-term solution.

Categories of Overexploitation

Overexploitation can be divided into two major categories: direct and indirect exploitation. Direct exploitation ranges from commercial activities such as logging operations or trade in endangered species, to subsistence hunting and fishing. Indirect exploitation encompasses the unintentional death of nontarget species, such as turtles and fish killed as by-catch in fishery operations. Both categories endanger large numbers of species around the world.

Direct, Commercial Overexploitation

While not all commercial ventures lead to overuse of resources, commercial exploitation is a major cause of overexploited resources. Natural resources are generally communal, and therefore vulnerable. In a system with communal resources, the cost of overexploitation is often borne by the whole community, not just the person using the resource, while the benefits go to the exploiter alone. It is in the best interest of individuals—in this case commercial operators—to overexploit communal resources until there is nothing left; this phenomenon has been dubbed the "Tragedy of the Commons." There are numerous examples of commercial overexploitation. We will consider three very important ones: overexploitation of marine fisheries, the wildlife trade, and logging.

Marine Fisheries

Oceans were once considered a limitless resource. This philosophy, coupled with a policy of open access to the oceans, set the stage for overexploitation. According to the Food and Agriculture Organization, the world's total capture fishery harvest reached 95 million tons in 2004, with marine catch accounting for 90.3 percent of the harvest. Among the major marine stocks exploited, 47–50 percent of species are considered fully exploited and are close to their maximum harvest, another 15 to 18 percent are overexploited and 9–10 percent of stocks are depleted (see Figure 4.6).

Major fisheries have collapsed around the world, from the Peruvian anchovy (*Engraulis ringens*) fishery in the 1970s to the cod fishery in Eastern North America in the 1990s. These collapses followed similar patterns. Initially these fisheries were so plentiful that they were judged impossible to overharvest. Harvesting systematically removed the largest, oldest fish from the populations. The largest fish are often the top predators, and removing them affects their prey and other predators. The oldest fish generally have the highest reproductive capacity and their loss exacer-

Figure 4.6 About 50 percent of marine fish stock are considered fully exploited, about 15–18 percent are considered overexploited, and another 9–10 percent of stocks are depleted (*National Oceanic and Atmospheric Administration [NOAA], Department of Commerce, Bob Williams, Photographer*)

bates already declining populations. Over time, boats had to travel farther and fish longer to harvest the same catch. At the same time, the average size of the fish caught began to decline substantially. In 1963, the average swordfish caught in the East Coast of North America was 250 pounds (113.6 kg); by 1996 this had dropped to 90 pounds (40.9 kg).

As one species becomes overexploited, fishing pressure simply shifts to another species—overharvested top predators are replaced with target species further down the food web. Between 1950 and 1994, there has been a gradual shift in the mean trophic level fished. In other words, as fish higher on the food web become depleted, fishermen fish for their prey species instead. Thus, fishermen are fishing fewer long-lived bottom fish that eat other fish (species higher on the tropic level), and more invertebrates and open water fish that eat plankton (species lower on the tropic level). This shift—termed "fishing down marine food webs"— has been most noticeable in the Northern Hemisphere and while it initially leads to an increase in catch, it is followed by declines. As we systematically remove the top predators and their prey from marine systems, we have put the oceans in a perilous state for recovery.

Recent research into historical and archeological evidence highlights the toll of overexploitation on many marine systems. The resulting impoverished state of these marine systems leaves them more susceptible to major disturbances (e.g., epidemic diseases, hurricanes, and climate change) and less productive for current and future human needs. For instance, in Caribbean coral reefs, populations of predatory and large herbivorous fish were overfished during the seventeenth to twentieth centuries. The loss of these fish made these reefs more susceptible to other threats. An introduced disease killed off most of the sea urchins (*Diadema antillarum*) in 1983 and 1984, removing the other major herbivore in the reef system. With the loss of these herbivores, corals of the Caribbean perished under overgrowth of macroalgae.

As in other cases of commercial overexploitation, technological advances have significantly contributed to overharvesting of marine fish. Engines, refrigeration, sonar, global positioning systems (GPS), and acoustic Doppler profilers have made it easier to locate, catch, and store fish, and allowed fishermen to fish farther from shore and for longer periods of time. New fishing gear has allowed us to harvest fish faster and in areas that were once inaccessible. For example, long lining, a technique where a line extending 100 kilometers is baited along its length with thousands of hooks, enables fisherman to catch the same amount of swordfish in just 3 days that would have been caught in 2 weeks by harpooning. New trawling techniques allow access to areas previously out of reach.

Regulations have been imposed to try to control the exploitation of fish with mixed success. In the 1970s, a 200-mile limit was imposed around the world's coastline to enable countries to regulate fishery harvest in their waters. Quotas on the number and size of fish caught, restrictions on the fishing gear, and limitations on the number of boats allowed into a fishery have been used to help control harvest rates.

Aquaculture was considered a solution to already overfished oceans. Unfortunately, the marine species farmed are often carnivores and require wild-caught fish as food. It takes 5 pounds of wild fish to raise 1 pound of farmed salmon. So rather than reducing harvest, aquaculture has placed a new burden on fisheries to supply fish meal.

Wildlife Trade

According to Interpol the illegal wildlife trade value is estimated at $20 billion a year, second only to drugs in value, and is frequently connected to other organized crime. TRAFFIC, an international organization established by the World Wildlife Fund and the World Conservation Union (IUCN), monitors the trade in wildlife and wildlife products. Based on declared import values, TRAFFIC estimates that the global wildlife trade is huge, with an annual turnover of billions of dollars and involving hundreds of millions of individual plants and animals. The Convention on the International Trade in Endangered Species of Wild Fauna and Flora (CITES) regulates international trade in some 30,000 species of plants and animals through a system of certificates and permits. A large proportion of the world's wildlife trade does not cross international borders, especially trade in medicinal plants, timber, wild meat, and fisheries (see Figure 4.7). The magnitude of the domestic trade for most wildlife species remains unknown.

Figure 4.7 Wildlife trade is a problem in many parts of the world for everything from pets to curios to food (*Snyder©CBC-AMNH*. Used by permission)

Commercial hunting of wildlife, prevalent across tropical Asia, Africa, and the Neotropics, is a growing problem. The wildlife trade has increased, as modern weapons replace traditional ones, and as new roads increase access to remote areas, and reduce transport times between the forest and markets. As humans colonize formerly remote regions, few places are immune to the effects of the wildlife trade. In some communities harvesting wild plants and animals provides additional income, whereas in other communities, these resources are irreplaceable for food.

Wildlife is a critically important resource meeting food and livelihood needs of human communities in many biodiversity-rich areas of the world. The combined economic value of wild meat from subsistence use, legal and illegal commercial trade contributes significantly to many local and national economies. Wild meat is a significant source of easily accessible animal protein (and often the only form of protein) for landless, rural peoples throughout Asia, Africa, and Latin America. In the Malaysian state of Sarawak on the island of Borneo, 67 percent of the meals of the indigenous group known as the Kelabits, contain wild meat, and it is their main source of protein. In Liberia, 75 percent of the country's meat is derived from wild animals. Wild meat is also of high economic value; it is an important source of livelihood for rural and urban communities for both subsistence use and commercial trade. The value of wild meat harvested in the Amazon Basin exceeds $175 million USD per year and in Côte d'Ivoire, the value is estimated to be $200 million USD.

It is increasingly difficult to distinguish subsistence hunting from commercial hunting, traditional from modern, and sport from necessity. These changes in hunting practice complicate the search for sustainable solutions. Fundamentally, rising human populations and consumption rates drive the transition from subsistence to commercial hunting.

Forest Exploitation

Trees are overexploited to supply timber, paper, and fuel wood markets. Forests are also exploited for their nontimber products (such as for nuts, rubber, dyes, cork, etc.); however, the processes for extracting these products typically leave the forest intact. Though nontimber products can be overharvested, if harvested sustainably, they also offer a potential tool for forest conservation.

Exacerbating these direct threats to forests is the conversion of many forests for agriculture and settlement. Today forest cover has shrunk to approximately half of its *potential extent*—that is the extent of coverage if there were no humans and based on the current climate—replaced by agriculture, grazing, and settlement. Primary forest of a significant size remains in only a few countries—notably the boreal forests of Northern Canada and Russia, the Amazon basin of Brazil, and Central Africa's Congo Forest.

The world's forests began declining thousands of years ago, with the expansion of farming and herding in the Middle East and Europe. More recently, rapid population growth, industrialization, and globalization are contributing to rapid deforestation in many tropical regions, and forest loss in Brazil and Indonesia exceeded 3.5 million hectares in 1995. In 2006–2007 forest loss in Brazil was at its lowest since the 1970s with 3,700 square miles cut for the year ending July 31. However, in the last 4 months of 2007, deforestation jumped to 2,300 square miles. While there is no question that forest loss and fragmentation is substantial, determining the exact rate of these losses globally is complex. Determining rates of forest loss at smaller, local scales is often easier, but they too can be controversial. Why do estimates of forest loss vary so much?

• differences in the classification of forest types;
• limited data, especially historical data and for certain parts of the world (developing countries) for some regions;
• limited verification of satellite data in the field; and
• poor or irregular government reporting.

These factors must all be kept in mind when examining data on the extent and rate of forest loss and fragmentation.

Besides the obvious loss of forest, logging often disturbs forests leaving behind an altered ecosystem. The Middle East and the Mediterranean suffered extensive forest loss thousands of years ago and the remaining forests are highly degraded even to this day. In Turkey, for example, selective logging of pine forests removed the tallest, straightest trees. Today's Turkish forests are consequently made up of smaller trees with more irregular shapes. The level of degradation caused by logging depends on its intensity: selective logging causes moderate disturbance, whereas heavy logging activity drastically modifies the forest environment from its original state. Roads built for logging increase access to remote areas for settlement and hunting, and also fragment existing forests increasing the risk of invasion by nonnative plants. Additionally, logging can make areas more vulnerable to fires. Logging waste, coupled with undergrowth and pioneer plants that colonize an area after it has been logged, serve as fuel for fires. Wildfires have become a major concern around the world, especially since the 1990s.

Forests are also exploited for their nontimber forest products. These represent a diversity of materials including food, medicine, spices, essential oils, resins, gums, latexes, tannin, dyes, and fibers. Unlike timber extraction, exploitation of nontimber forest products is often less destructive and may serve as an economic incentive to conserve forests; however, they can also be overharvested. Also, increased understanding of the value of forest ecosystems in conserving soil and water, and regulating climate may further assist conservation efforts and reduce the continued exploitation of remaining forests.

Indirect Overexploitation

Nontarget species are inadvertently exploited when target species are harvested for commercial use. Though occasionally these unwanted, nontarget species are sold, often they are simply discarded. For instance, many fishing methods are not selective, catching other species besides the intended one. This indirect harvest—or "by-catch"—can be substantial. Total global by-catch is estimated at 16–40 million tons per year (Alverson et al., 1994), and impacts many species particularly marine mammals, turtles, and seabirds. The worst offenders are shrimp and bottom trawlers. Shrimp harvest has one of the highest by-catches of any fishery; for every pound of shrimp harvested, about 5 pounds of by-catch is caught and wasted, though sometimes by-catch can reach 14 pounds. Long-lining for tuna and swordfish also have significant by-catch of sharks, sea turtles, and marlin. Tuna fishing practices have since been modified after public outcry because of the high number of dolphins caught in the nets. However, there is still concern that some "dolphin-friendly" methods still lead to high levels of stress. While the dolphins survive, the stress seems to greatly impact their reproductive rates.

More efficient trawling methods took their toll on the barndoor skate (*Raja laevis*), which is now on the brink of extinction. In fact, its disappearance nearly went unnoticed, until surveys for halibut uncovered a refuge for the species and more detailed analysis was conducted. Barndoor skates are the largest skates in the Northeastern Atlantic reaching over 1 meter in length. They are also slow-growing, maturing at around 11 years of age, and then only producing a small number of eggs. These characteristics make them especially susceptible to overharvesting. These skates were once common along the eastern coast of North America, especially in the Gulf of Maine and off Southern New England coast. In the 1950s and 1960s, they were an abundant and common by-catch when other species, like cod, were harvested, representing about 10 percent of the catch in trawl nets. Now it is a rarity and is a candidate for listing as an endangered species, as a consequence of overharvesting.

On coral reefs, fishing with dynamite or cyanide results in substantial by-catch. These destructive fishing techniques, though illegal, are still widely used, as enforcement in many countries is limited. Dynamite fishing not only kills the fish and invertebrates nearby the target species, but also destroys the physical structure of the reef, causing long-term damage to the entire community. To fish with cyanide, divers locate the fish and squeeze a bottle filled with cyanide into the water, temporarily stunning the fish. The cyanide can damage the reefs and nearby fish. The target fish are transported live either to restaurants or for the aquarium trade; many die within a few days of capture or during transport. An estimated 70 to 90 percent of fish caught by this method die before they are even sold.

Similarly, the wildlife trade offers yet another example of indirect overexploitation. Many animals die during capture and shipment, essentially bolstering the numbers that need to be captured in order to adequately supply the market destination. Mortality rates can reach up to 60 to 70 percent for some birds and reptiles. So to ensure 10 parrots are delivered, they will catch 13 or 14 parrots, expecting three or four to die in transit.

POLLUTION

Every day, thousands of pollutants are discharged into the environment. Many are unregulated, and their lingering presence threatens biodiversity, affecting individual species or degrading entire ecosystems, and human health.

What Makes Something a Pollutant?

Pollutants resist categorization because of their varied forms and effects. Some directly toxify the environment, such as lead or PCBs. Other pollutants, such as runoff from fertilized lawns, are nontoxic but harm aquatic systems by causing excessive plant growth. Noise and light pollution threaten species by disrupting their behavior. Pollutants are also classified by the environment they affect, such as air, water, and soil pollution, though many pollutants cycle through all these environments at some stage, entering the air and ending up in water or soil. Classification of pollutants may also derive from where they enter the environment: a "point source" pollutant enters at a discrete location and is nonmobile,

Figure 4.8 It is easier to regulate "non-point" source pollution, such as from a factory, than "non-point" source pollution (©*Ryan Luci*. Used by permission)

such as effluent from a sewage treatment plant, whereas a "nonpoint" source pollutant enters from many locations or is mobile, such as surface runoff into the coastal zone from cars (oil), or lawns (fertilizers and pesticides). Typically, it has been easier to regulate "point source" pollutants (see Figure 4.8).

Though difficult to characterize, some typical characteristics of pollutants included their tendency to persist in the environment. Because of this, even after a pollutant is banned, its legacy remains. Pollutants are often widespread and can be transported over large distances. Many accumulate in animal tissues or interfere with vital processes such as reproduction or immunity. Some are toxic even at tiny concentrations and at the extreme will kill an animal. Pollutants can also substantially alter entire ecosystems.

Because so many pollutants enter our air, water, and soil, it would be difficult to cover them all. Here we examine only some of the leading pollutants of our environment, including toxic contaminants, organic biostimulants, solid waste, noise, and light pollution.

Toxic Contaminants

Toxic contaminants include trace metals (e.g., cadmium, copper, lead, and mercury), biocides/pesticides (e.g., DDT, TBT [tributyl tin], industrial organic chemicals (e.g., PCBs, tetrachlorobenzene), and byproducts of industrial processes and combustion (e.g., polycyclic aromatic hydrocarbons [PAHs] and dioxins). They can be lethal or interfere with an organism's immune, endocrine, and reproductive systems.

Chlorinated hydrocarbons, such as the insecticide DDT (dichloro-diphenyl-trichloro ethane) and PCB (polychlorobenzene), are renowned for their toxic effects on the environment. A particularly troubling characteristic of these pollutants is their ability to persist over long time frames and spread over large areas. When DDT was introduced in the 1940s, it was a marvel; it was cheaper and more effective than any other insecticide. However, its effectiveness came at a price. During the 1950s and 1960s, populations of predatory birds in North America—in particular, those that ate fish, including eagles, pelicans, and ospreys—declined rapidly. Analysis of the birds revealed that DDT in their bodies was a million times more concentrated than that of the water where they lived. This discovery led to the concept of bioaccumulation, that is, animals higher up the food chain concentrate contaminants in their bodies.

Why is DDT such a powerful toxin? First, it cannot be broken down by the body and is fat soluble, allowing it to accumulate in animal tissue. Second, DDT interferes with calcium deposition in eggs; thus birds were laying thin, fragile eggs that often broke during incubation. Because it affected the birds' reproduction, it had an immediate and powerful effect on populations.

DDT also disperses readily in the atmosphere and has even been found in animals in the Arctic and Antarctic. Even though DDT was banned in 1972 in the United States, it continues to persist in the environment. One notable place is in Palos Verdes, off the coast of Los Angeles, California. DDT manufacturers discharged their excess DDT into the sewage system and eventually the ocean—over 1,700 tons contaminates a 20 square mile area extending from the sewage outfall pipe. The same site is also contaminated with PCBs. Manufacturers continue to export DDT and other pesticides that are banned in the United States to developing countries. However, the Stockholm Convention on Persistent Organic Pollutants, signed in 2001 and ratified in 2004 by 128 countries, has banned the use of DDT except for the control of mosquitoes in areas prone to malaria, as there are few affordable alternatives. The Convention also bans another eight chemicals that make up what is known as the "Dirty Dozen." Furthermore, the Convention is designed so that other chemicals can be added to the list in the future.

Endocrine Disrupters

Many pesticides, including DDT and PCB, as well as DBCP (di-bromochloropropane), DDE (dichloro-diphenyl-dichloro ethylene), kepone, heptachlor, chlordane, dieldrin, mirex, lindane, toxaphene, dioxins, Bisphenol-A, and phthalates, are *endocrine disrupters*—chemicals that mimic or inhibit the effects of hormones. Most of these chemicals are long-lived compounds and bioaccumulate. The toxin tributyltin (TBT) used in antifouling paint on ships interferes with sexual development in some mollusks; for example, females exposed to TBT will develop male organs, even at concentrations of 10 parts per trillion. Declines in marine snail populations have been found along the coasts of North America and Europe, due to heavily contamination with TBT. International Maritime Organization established a Convention with a ban on TBT paints on ships effective as early as 2003, depending on the size of the ship; the ban has forced companies that make antifouling paints to research alternatives. Many have chosen copper-based paints as alternatives though these also cause pollution. Companies are still looking for nontoxic options that are effective and not too expensive.

Atrazine, a common weedkiller used heavily on corn crops and the most widely used herbicide in the United States, pervades the environment, contaminating rain and snowwater runoff, and groundwater. Atrazine is particularly popular as it enables farmers to use "no till" farming methods, which reduces soil erosion. According to the Environmental Protection Agency, 77 million pounds of atrazine are applied to U.S. crops each year. Atrazine, even at low doses, appears to affect frog development. One study found that 20 percent of frogs exposed to doses of just 0.1 parts per billion (well below the limit allowed for drinking water), developed abnormal reproductive parts, such as multiple sex organs or both male and female organs. At slightly higher doses of one part per billion, 90 percent of males lacked vocal chords, which are essential for attracting mates. Atrazine appears to affect the production of the enzyme aromatase, which converts the male hormone testosterone into the female hormone estrogen. Studies of frogs near farms in Vermont also found they were two times as likely to be deformed and the culprits appeared to be atrazine along with metolachlor, another herbicide. While atrazine is banned in the European Union, a very similar compound is still widely used there. EPA recertified atrazine in 2003; and in 2007 an independent scientific advisory panel solicited by the EPA held that atrazine did not affect gonad development. However, the use of atrazine continues to be controversial.

Acid Rain

Sulfur and nitrogen oxides are released into the air when wood or fossil fuels (such as coal in power plants, or oil in vehicles) are burned.

These combine with water in the atmosphere to create sulfuric and nitric acid, which eventually falls to earth as "acid rain" (rain with a pH less than 5.0). These pollutants also create "smog" in urban areas. Because of prevailing wind patterns and geological characteristics, certain regions (including the Northeast United States, Canada, and Northern and Central Europe) have been especially affected by these pollutants. Some soils and rock types neutralize or buffer the acid. For example, calcium carbonate in limestone acts as a natural buffer, reducing the damaging effects of acid rain. On the other hand, areas with granite and quartz tend to have low buffering capacity and acidify quickly. Freshwater lakes in these areas are particularly susceptible to acid rain. Initially, the changes affect only some species of invertebrates, but with increasing acidity fewer and fewer species survive until eventually the lake is dead, despite appearing "clear," and "clean." This phenomenon has been widespread in lakes in New York's Adirondack Mountains, northern Sweden, and Canada. Acid rain also dissolves harmful metals, like mercury, from surrounding rocks, which plants and animals then absorb. On land, pollution by acid rain and other air pollutants (ozone) tends to affect plants more than animals. Lichens, bryophytes, and fungi suffer the most. Declines may be due to acidification of the soil, direct toxicity, or competition from more resistant species. Animals, like otter and deer, tend to be indirectly affected by acid rain pollution due to changes in abundance of their prey or from the bioaccumulation of mercury in their tissue, which is released at higher acidities.

Organic Pollutants or Biostimulants

Organic pollutants or biostimulants, primarily from agricultural fertilizers and sewage waste, have a major impact on aquatic environments. When these excess nutrients enter aquatic systems they stimulate plant growth. Rapid phytoplankton growth or algal blooms create diverse problems. Plant growth is so rapid that animals can't consume them, the surplus phytoplankton then falls to the seafloor and decomposes. This decomposition depletes oxygen, creating hypoxic (low oxygen) or even anoxic (no oxygen) environments, where few organisms can survive. Large concentrations of algae also reduce water clarity, preventing light from reaching the bottom and reducing the growth of seagrasses. Changing phytoplankton communities also affects shellfish populations who consume them. A long-term increase in excess nutrients into an ecosystem is known as eutrophication. Over 50 percent of the estuaries along the coast of the United States are affected by eutrophication, some severely, such as the Mississippi River delta, Chesapeake Bay, and the Long Island Sound. Eutrophication is a worldwide phenomenon affecting coastal areas, from Europe to Asia.

Aquaculture operations also produce organic waste through uneaten food, feces and urine, and dead fish. While still a minor organic pollutant, it can have a major local impact. Areas with offshore salmon pen farming (like L'Etang Inlet, New Brunswick, Canada, and Puget Sound, Washington, United States), have significant nitrogen and phosphorous inputs due to aquaculture. Directly beneath the pens, there is often an anoxic area that extends 100 to 500 feet (30 to 150 meters) from the caged area. Effluent from freshwater pond aquaculture (such as that used for shrimp and catfish) also contaminates nearby waterways when waste is released.

Besides releasing organic nutrients, aquaculture releases a mix of chemical and biological pollutants. Antibiotics, parasiticides, pesticides, hormones, anesthetics, pigments, minerals, and vitamins are added to the feed for various types of pen and pond aquaculture systems. Especially in pen aquaculture, which is completely open to the surrounding water, uneaten food enters the water where it can contaminate wild species. Similarly, escaped fish are a form of biological contaminant. Farm-raised fish have been bred with certain traits; when they escape they can reproduce with and alter the wild population. New methods of raising fish in aquaculture are being tested in order to minimize these negative effects on the environment.

AMPHIBIAN DECLINES

Recent global assessments indicate that 32 percent of the world's 6,000 amphibians are threatened. A number of culprits appear to be contributing to this decline, among them disease, and increased nitrogen from agricultural fertilizer. Since students on a school field trip in southern Minnesota in 1995 found a large number of malformed frogs, the extent of amphibian malformations has become a growing concern. The North America Reporting Center for Amphibian Malformations (NARCAM) was developed to coordinate and monitor the occurrence of amphibian abnormalities. As of 2002, there were 944 verified reports of malformations involving 52 species in 46 states and four Canadian provinces. While there is still uncertainty about the level of malformations that occur naturally, there is growing evidence that amphibian malformations are an unusual and recent occurrence in parts of North America.

Malformations usually represent an error that occurs during an amphibian's development. Currently field and lab analyses point to three main causes of abnormal development: ultraviolet (UV-B) radiation, chemicals, and parasites, though other agents may exist but have not yet been identified. These may cause abnormalities by interfering with signaling molecules, such as retinoic acid, during limb development.

The formation of multiple limbs is also highly linked to infection by larvae of the trematode flatworm Ribeiroia. The parasite's life cycle is complex. After being released from their hosts (likely birds), Ribeiroia eggs are picked up by snails. They then develop into swimming larvae that eventually rest on the developing pelvis and hind limb buds of frog tadpoles, where they become cysts. Increased agricultural fertilizer and pesticide runoff (made up of nitrates and nitrites) may increase the parasite's effect on amphibians. Several amphibians are affected by fairly low levels of nitrates and nitrites during the larval stage. In addition to these direct effects on amphibians, nitrates and nitrites increase algal growth and potentially boost populations of snails, the trematode host. This in turn leads to increased deformity rates in nearby amphibian populations. Studies in Vermont, where Ribeiroia is not found, revealed that frogs near farms were twice as likely to be deformed and the type of deformations found were distinct from those caused by the flatworm. This suggests that pollution from agriculture alone may be enough to cause deformities.

Our ecosystems are fundamentally altered by pollution. It was long thought that it was normal for temperate forests to lose nitrogen into soil and stream waters in inorganic forms, such as nitrate and ammonium; however, studies of ancient unpolluted temperate forests in Chile and Argentina reveal this may be an artifact of pollution. South American forests are dominated by the release of dissolved organic nitrogen. Their North American counterparts (in this study the Smokey Mountains of Tennessee and Tionesta National Forest in Pennsylvania were analyzed) release high levels of inorganic nitrogen. How nitrogen cycles through North American temperate forests appears to be influenced by excessive fertilizer use and nitrogen deposition from acid rain.

Solid Waste

Solid waste is generated from household and industrial sources, and includes everything from food to plastics. Most solid waste ends up in landfills. Landfills take up space and if not properly contained leach toxins—termed leachate—into the soil and groundwater. In countries with limited space, solid waste is burned at high temperatures. But incineration is expensive, creates very hazardous ash, and pollutes the air with toxic chemicals. Solid waste can be minimized through recycling and composting. Certain materials, namely metals, glass, and paper, are easier to recycle. Recycling reduces waste and creates economically valuable products. Composting of organic materials, such as food and paper, produces a natural fertilizer, and is an effective way to reduce solid waste and methane emissions.

Mining Waste

When people think of solid waste, most think of their own garbage, but mining produces huge amounts of solid waste. In the United States, mining produces over 1.7 billion tons of waste compared to the 180 million tons produced by all municipalities combined. Extraction of minerals, coal, and oil destroys and fragments habitats, and is very polluting. In the worst instances, mineral, coal, and oil extraction cause catastrophic spills.

Open-pit mining is an extremely wasteful process. Metals, like gold or copper, and mineral substances, such as coal, are extracted from ore found close to the surface. Most ore contains only small amounts of the target metal; the remaining excavated rock is wasted. The amount of waste depends on the metal and the region being mined, but typically it is huge. Three tons of ore are needed to produce enough gold for just one ring. Copper mining is similarly wasteful—for every ton extracted, 99 tons of waste rock is produced. New technologies to extract and process minerals found only at low concentrations in ore are increasing the waste produced by the mining industry and enabling new areas to be exploited.

Much mining waste is hazardous, and the excessive quantities produced go on to pollute the environment with heavy metals, acid-producing sulphides, and other contaminants. Additional waste, known as tailings, is also produced during processing. Tailings are another toxic by-product of the mining process. They are finely ground material made up of heavy metals and chemicals, like cyanide, arsenic, and sulphuric acid. Mining waste and tailings are stored in special containment ponds near the mining site. Pollutants often leach from these sites into soils, groundwater, and nearby lakes and streams. If these sites are not well-maintained, disasters can occur. One example from southwestern Spain in 1998, a mining accident released 5 billion liters of toxic sludge into the Guadalquivir River. Contamination spread over a huge area downstream damaging the wetlands of Coto Doñana and the Doñana National Park.

— ✐ —

THE LEGACY OF MINING

Mining leaves a long legacy on the land. While they have long stopped mining gold in California, historic mining has contaminated many of the state's rivers with heavy metals, like mercury. Water was used initially to break down the gold containing ore. To extract gold from ore, mercury was often added, and as ore is passed through a sluice it is mechanically extracted with the heavier gold attached to mercury separated from other materials. Mercury is dense, so once attached to gold, the particles tend to sink allowing them to be separated out easily. However, in the process mercury was often lost to the environment contaminating water and sediment. An estimated 10 million pounds of mercury is lost during processing, about 80 to 90 percent of that in mines in the Sierra Nevada Mountains. Mercury contamination is a problem in many rivers and lakes of northern California, where fish advisories are in place. Mercury contamination is a widespread problem, not only due to mining but also because it is a common air pollutant; over 40 states issue advisories about fish consumption.

One of the most wasteful methods of mining is called "mountain top removal." An entire mountain peak is removed to expose the underlying coal; the rock and earth removed is then dumped into surrounding valleys and streams. The end result is an extremely altered landscape with huge impacts on habitats and wildlife, stretching well beyond the site of the mine. This is a critical issue in the Appalachian region of the United States. There have been efforts to stop this practice by employing the Federal Clean Water Act. In January 2008, Massey Energy, one of the largest coal mining companies in the world, paid $20 million in fines due to 4,600 violations of the Clean Water Act in West Virginia and Kentucky. They will also have to take measures to prevent future violations.

Plastics in the Ocean

Waste originating on land can wash into storm drains after a heavy rain, be carried out to sea from beaches, is dumped by ships, or enters the ocean from the air, and all eventually pollute the marine environment. Floating debris (plastics, balloons, etc.) even reaches the world's remotest islands, miles from human settlement. Plastics and fishing gear threaten many marine species. Sea turtles appear to confuse plastic bags with jellyfish, one of the main prey items for some species. The plastic blocks their digestive track, killing the turtles. Studies of stranded sea turtles off the coast of Brazil found that the most common debris ingested were transparent white plastic bags; the turtles also showed evidence of damage from fishing activities on their carapaces. Recent studies of 30 remote island sites around the world revealed that floating marine debris is mostly made up of plastics. In addition to harming the marine mammals that swallow them, these plastics act as rafts spreading invasive species, like barnacles and mollusks, around the globe. Lost or discarded fishing gear, another major source of marine pollution, can continue to fish for many years, entangling turtles, whales, seals, seabirds, and fish. Gear also damages the reef and benthic habitats that support marine life.

Plastic that doesn't end up being swallowed by some unfortunate animal ends up leaching chemicals into the environment as it breaks down. In 2004, scientists in the United Kingdom discovered that there were microscopic levels of plastic in sand taken from the country's beaches. Plastic and other synthetic materials, like acrylic and polyester, get deposited in sediment from the breakdown of garbage seen floating around in the world's oceans and other waterways. When small animals, like barnacles, ingest sand, the plastic shows up in their digestive tract within a few days. Scientists are now studying whether toxic chemicals adhere to these tiny bits of plastic and what effects this could have as the plastic moves up through the food chain.

Noise Pollution

Transportation (cars, trains, airplanes, and ships) and industry (construction or manufacturing) are the leading sources of noise pollution. Animals rely on hearing to communicate, avoid predators, and obtain food. To avoid noise, wildlife may alter their behavior, leaving critical habitat or forage areas. Noise can cause hearing loss, interferes with communication, and long-term exposure may have physiological effects brought on by faster heart rates and metabolism.

Many studies have examined the effects of noise on wildlife. Responses vary with the type of noise and the species. Magnificent frigate birds (*Fregata magnificens*) in the Florida Keys appear to be disturbed by low-flying aircraft at their nesting sites. Birds flushed from their nests when they hear a noise may even break their eggs or injure the young. Caribou calves (*Rangifer taranuis*) exposed to overflights also suffered higher mortality rates than unexposed populations. Many desert animals have acute hearing and depend on it for hunting. Colorado Desert fringe-toed lizard (*Uma notata*) and the endangered Desert Kangaroo rat (*Dipodomys deserti*) experience hearing loss due to motorcycle noise. Desert bighorn sheep (*Ovis canadensis nelsoni*) in the Grand Canyon were particularly sensitive to helicopter passes in the winter, evidently because they grazed at higher elevations than during the spring and were closer to the sound.

Noise from shipping, fishing, recreational boating, dredging, military activities, or oil exploration disturbs marine animals. Whales and dolphins, which rely on sound for communication and navigation, appear particularly affected. Whales startled by noise (especially at low frequencies) may dive suddenly, swim faster, or change their vocalizations. At the extreme, noise may even kill them. In March 2000, nine Cuvier's beak whales (*Ziphius cavirostris*), three Blainville's beaked whales (*Mesoplodon densirostris*), two unidentified beaked whales, two Minke whales (*Balaenoptera acutorostrata*), and one spotted dolphin (*Stenella frontalis*) were stranded in the Bahamas, some bleeding from their ears, and at least seven of them died. According to the U.S. Navy and the National Marine Fisheries Service (2001), sonar testing in the area was linked to the strandings. The marine mammals were confined to a narrow channel off the coast, which tend to amplify sound during calm conditions. Recent studies also show that whales, like human divers, are susceptible to diving illnesses. Sonar and noise from explosives used in military activities or during oil exploration can cause marine mammals to dive deeper. On long deep dives, more nitrogen enters the blood from the lungs as bubbles; too much nitrogen in the bloodstream can kill an animal. The U.S. Navy is currently barred by a court order from using sonar within 12 miles of the coast or if whales are sighted within 2 kilometers; a 2008 California district court upheld the order.

Light Pollution

Satellite images of the planet at night reveal dramatically the extent of light pollution on the planet. Urbanized areas, and most noticeably the coastlines, are completely "lit up." Light pollution impacts many nocturnal species.

The effect of light pollution on nesting turtles and hatchlings is well documented. Turtles normally use moonlight to guide them back to the ocean, but instead walk toward brighter artificial lights on land. At night during foggy weather, when visibility is low, migrating birds can become disoriented by radio tower lights (especially from towers higher than 200 meters). In the United States there are over 40,000 towers and where studies have been conducted, mortality rates have ranged from 375 to 3,285 birds at a tower per year. At one tower a thousand birds have been killed on one night.

It is common knowledge that lights attract moths and night-flying insects, but few realize this may affect their populations. Declines in moth populations may be linked to the effect of artificial lights on their reproduction. The energy moths spend attracted to artificial lights may prevent them from finding mates or good places to lay their eggs. Artificial light may even impact plants; plants whose reproduction is controlled by set night-day lengths may not flower due to artificial lighting.

GLOBAL CLIMATE CHANGE

Global climate change is expected to impact plants and animals worldwide. The exact changes will depend largely on the amount and rate at which the world's climate warms. New studies and improved models are increasing our understanding of global climate change. Over the past 100 years (1906–2005), the global average surface temperature has increased 0.74°C ± –0.18°C, according to the International Panel on Climate Change 2007 Report. The warming trend over the last 50 years (0.13 ± –0.03°C) per decade) is nearly twice that for the last 100 years. Eleven of the last 12 years (1995–2006) were among the warmest years since 1850. Though the average global surface temperature has increased by 0.76°C ± –0.19°C since 1850, there has been great regional variation, with some regions experiencing much larger increases and others smaller or no increases in temperature. The average minimum temperature has also increased at a faster rate than the average maximum temperature. While it is difficult to examine temperature change on longer time scales, recent studies indicate that the temperature increase in the Northern Hemisphere is likely the largest of any century in the past 1,300 years; unfortunately, fewer data are available to predict this for the Southern Hemisphere.

In addition to increasing temperatures, there are other indications that the earth's climate is warming. Satellite data reveals that snow cover has declined 10 percent since the 1960s. During the twentieth century, there has been a continued retreat of the world's mountain glaciers, and at mid- and high-latitudes in the Northern Hemisphere the time that lakes and rivers remain frozen has decreased by an average of 2 weeks. In late summer through early autumn, the thickness of Arctic sea ice appears to have thinned by 40 percent, while the extent of sea ice has shrunk by 2.7 percent per decade. Sea level rose 1.8 mm/year since 1961 and 3.1 mm/year since 1993, largely due to thermal expansion (water expands at higher temperatures), melting glaciers and ice caps, and polar ice sheets. Between 1900 and 2000, precipitation has increased significantly in eastern North and South America, northern Europe and northern and central Asia, although it has decreased in the Sahel, the Mediterranean, southern Africa, and parts of southern Asia. Increasing temperatures have been accompanied by shifts in the period of seasons with earlier springs and longer autumns. As a result, over the last 40 years, the growing season in the Northern Hemisphere has lengthened from 1 to 4 days per decade.

***

DISAPPEARING ICE SHEETS

The world's two largest ice sheets—Antarctica and Greenland—are also melting. While there is still uncertainty about the extent and rate at which these ice sheets are melting, by some estimates this could lead to catastrophic sea level rise up to 20 feet. The southern part of Greenland lost 54 cubic miles of ice in 2005, twice the amount lost 10 years earlier. Over the last 15 years, the time of the melting season has gotten longer, starting earlier and ending later.

The much larger Antarctic ice sheet—with more than 90 percent of the word's ice—while more stable than its northern counterpart also shows signs of increased melting. Seventy-five percent of 400 mountain glaciers on the Antarctic appear to be retreating. Additional studies of 300 mountain glaciers show an increase of 12 percent in flow between 1993 and 2003 (see Figure 4.9). In March 2002, the Larsen B ice shelf collapsed into icebergs; after this collapse other ice floes sped up in the area. The West Antarctic Ice Sheet is also calving off icebergs. If it were to melt it could raise sea level as much as 20 feet. There is still controversy over the contribution of the melting of these ice sheets to sea level rise largely due to uncertainty as to the speed at which the ice sheets might melt and the magnitude of the change.

***

Current Global Circulation Models predict that in the future, the globally averaged surface temperature will increase by 1.8 to 4.0°C between 1990 and 2100 (ranges are 1.1–2.9°C and 2.4–6.4°C), while sea level will rise between 0.09 to 0.88 meters (3.5 to 35 inches). These averages are for the entire planet and a large degree of regional variation is anticipated. Notably, climate change is expected to have a disproportionate effect at higher latitudes, which will have larger temperature increases.

What Does Climate Change Mean for the World's Ecosystems and Species?

Climate is central to the geographic distribution of the world's vegetation types and animal species. As the climate warms, we expect to see shifts in vegetation patterns and species distributions. These shifts may fundamentally alter ecosystem composition and function. A 3-degree increase in temperature corresponds to a 500 meters (547 yards) altitudinal shift or a 250 kilometers (155 miles) latitudinal shift—in other words, species found at 500 meters elevation will now be able to migrate up to 1,000 meters or be able to expand their range northward by 250

Figure 4.9 Disappearing mountain glaciers and the increased melting of the world's major ice sheets are signs of the warming climate (*Domroese©CBC-AMNH.* Used by permission)

kilometers. This speed of change is similar to the change in climate during the Pleistocene era, which was too rapid for many species to adapt. Alpine species may disappear entirely as they are pushed to their distributional limits. Those species that can adapt fast enough may encounter other barriers, such as human development, which hinders their ability to adjust to climate change.

Of the 29,000 observational data in 75 studies, more than 89 percent show changes in biological and physical systems consistent with climate change. For example, studies have found that European butterflies were changing their ranges in response to climate change. Of the 35 species examined, 63 percent had shifted their range north by 35–240 kilometers (21–149 miles). Only 3 percent had shifted their range southward. This shift reflects changes in colonization and extinction rates at the boundaries of the species range. Europe has warmed an average of 0.8°C this century, reflecting a northward shift in climate of about 120 kilometers (75 miles). Warming temperatures may also allow some insect pests to widen their range, like mosquitoes that transmit malaria and dengue fever.

Coral Bleaching

Species that live close to their temperature limit are particularly vulnerable to climate change. Corals flourish at temperatures between 16°C and 25°C. Excessive temperatures stress corals and at the extreme lead to "bleaching events." Bleaching is when corals expel their symbiotic zooxanthellae—unicellular algae that photosynthesize and are found within corals. Bleaching was once considered a rare, isolated event and the corals often recovered. During the 1980s, however, large-scale bleaching events caused extreme loss of coral, and since then bleaching has occurred somewhere in the world on an annual basis. The 1997–1998 bleaching was the most severe and widespread ever observed, affecting reefs in the Pacific and Indian Oceans, Red Sea, Persian Gulf, and the Caribbean, and destroyed about 16 percent of the world's reefs. Bleaching is usually confined to the surface areas (depths less than 15 m or 16.4 yards), in this instance the damage extended to depths of 50 meters (or 54.7 yards). The Great Barrier Reef experienced bleaching events in 1998, 2002, and 2006.

Alpine Species

Similarly, species and communities that live at high elevations and are adapted to the climatic conditions in those regions are expected to be heavily, and some of them irreversibly, impacted by climate change. Studies in Colombia indicate that as temperatures rise and glaciers recede, high altitude mountain ecosystems, such as the paramo, will shift upward with ecosystems at the highest altitudes significantly reduced in area or even disappearing entirely. Studies in other mountain regions have revealed similar patterns as a result of climate warming.

Migratory Species

Migratory species are also vulnerable to climate change. Seabirds time their migrations carefully to take advantage of prey resources along their route, such as the spawning of horseshoe crabs or krill. Warming temperatures alter this timing and may cause a species to miss these key resources on route. A long-term study in the American southwest has revealed a trend toward earlier breeding in the Mexican jay (*Aplelocoma ultramarine*). Similar patterns have been observed in the United Kingdom. It is unknown what effect this will have on their reproductive success.

Arctic and Antarctic Ecosystems

At higher latitudes, increasing temperatures appear to be altering the environment and affecting ecosystem function. In Canada's Hudson Bay, the weight and reproductive condition of polar bears (*Ursus maritimus*) have been declining since the early 1980s. At the same time, rising spring temperatures have led to earlier ice breakups. Polar bears need solid ice to hunt effectively, and in the spring, their main prey are young, ringed seal pups. Before the ice thaws in the spring, polar bears must gain sufficient weight to last through the fasting period in the summer, when they are unable to catch prey on the open water.

Coastal Ecosystems

Sea level rise can have dramatic consequences for coastal environments. Rising sea levels are due to a combination of thermal expansion and loss of glaciers in mountainous areas, and potentially the loss of polar ice sheets, though there is still much uncertainty about how ice sheets will respond to climate change. As with temperature change, sea level rise is expected to vary regionally. Low-lying coastal areas and small island states are particularly vulnerable to sea level change, because it increases the risk of coastal flooding and the impact of storm surge. Other consequences of sea level rise are the loss of beach, wetland, and mangrove habitats.

Endangered Ecosystems and Species

Finally, climate change may exacerbate pressures on already endangered systems. Species that are confined to a small fragmented habitat are particularly at risk to climate change as these species will not be able to migrate as vegetation and habitats shift. For example, wetlands have the potential to migrate landward as sea level rises, but this migration is seriously hampered by coastal development. Despite the complications in modeling ecosystem response to climate change, a study by the World Wildlife Fund (2000) undertook this challenge, analyzing the projected impacts of climate change across the globe with surprising results. Notably significant shifts in the distribution of vegetation are expected and very high migration rates of vegetation exceeding 1,000 meters per year would be needed for adaptation to climate warming.

Synergistic Effects

An important and often overlooked issue is that many threats to biodiversity acting together are often far more potent than the sum of their individual impacts. As such, they can be highly synergistic in effect. One example can illustrate. Around the world, temperate estuaries have been permanently changed by humans. Estuaries, like the Chesapeake Bay off the coast of Maryland and Virginia, are severely affected by nutrient pollution from agriculture and sewage runoff. These excess nutrients cause phytoplankton blooms, some of them toxic, which in turn are decreasing or eliminating oxygen from the bottom sediments, making them inhospitable to life. Historical analysis of the sediments conducted by Dr. Jeremy Jackson revealed that as early as the late eighteenth century, human settlement in the watershed was affecting nutrient loads to the estuary and consequently the type of phytoplankton that was growing. However, sediments were still not experiencing low oxygen conditions, because the bay also had acres of oysters. Oysters (*Crassostrea virginica*) could filter the entire Bay in a matter of days, removing the excess phytoplankton and maintaining oxygen levels. But humans then began harvesting oysters at increasing rates until the Bay was nearly depleted of oysters by the 1930s. Without oysters to control the impacts of excess nutrients from the land, the system collapsed and the Chesapeake is now substantially and possibly irreversibly altered. Chesapeake Bay is not the only place where multiple disturbances brought about the collapse of an entire ecosystem. For instance, the Hawaiian Islands harbor one of the Earth's most spectacular biotas, but also one of the most fragile and endangered. Exotic species introductions, in combination with habitat disturbance by humans, have transformed more than 90 percent of the natural areas in Hawaii and led to countless extinctions. Finally, in the Amazon region, the water lost from plants through evapotranspiration is believed to contribute 50 percent of the annual rainfall. Deforestation reduces evapotranspiration rates, leading to decreased rainfall, and subsequently increases the area's vulnerability to fire. Fire can quickly burn acres of forest. Deforestation thus leads to additional forest loss through its indirect affect on the climate.

Now more than ever, we are realizing that all life processes on earth are interconnected. Humans are affecting not only the species that will go extinct today, but also the possibility of what will evolve in the future. Yet the future of earth's biodiversity also depends on us. Only by understanding our impact on biodiversity and its importance to human survival, can we discover how to conserve it.

5

CONSERVING BIODIVERSITY: STRATEGIES AND SOLUTIONS

New and innovative ways of conserving biodiversity are emerging, and people are increasingly considering how they might live in harmony with nature. This chapter looks at how biodiversity is currently protected and suggests innovative strategies to conserve biodiversity. We include a list of the small changes that people can make to their daily life to make it more sustainable and reduce their impact on biodiversity. We also examine some of the new alternatives being implemented to create sustainable communities and businesses.

INTRODUCTION

Human activities are the principal driver behind many threats to biodiversity. As implicated as we are in biodiversity's demise, people can also conserve ecosystems and the living things that inhabit them. The survival of all life on earth—including the survival of humans—depends on a healthy environment.

Whether as individuals, or collectively, we all make decisions that affect biodiversity. Should a person eat swordfish? Should a village sell the timber from their forest or should they conserve it for its nontimber products and other benefits (e.g., fruit and nuts, animal products, medicinal plants, and shade)? Should a city allow development or protect open space? Should a company or university renovate its buildings or buy products that are sustainably harvested? Should environmental and biodiversity issues influence our decisions in the voting booth?

Just as the threats to biodiversity occur at multiple levels, so do the solutions to conserve it. International organizations, nations, nongovernmental organizations, academic institutions, local grassroots groups and, most importantly, individuals, can all adopt more sustainable lifestyles and make changes that will conserve biodiversity in the long run. As individuals, our power lies in our everyday actions through which we either contribute to the problem or help solve it. What follows are some of the most urgent and promising steps that we must take to conserve the world's biodiversity.

A BRIEF HISTORY OF THE CONSERVATION MOVEMENT

Though the "conservation movement" is often considered a recent phenomenon, conservation activities pepper historical accounts across cultures, regions, and through time. Even in ancient civilizations as far back as 3,000 years, and in places as distant as China and South America, people had set aside land for the protection of plants and animals. For example, some of the first recorded laws protecting forests were decreed in the ancient Mesopotamian city of Ur in 2700 BC.

Royalty were responsible for many early conservation efforts, setting aside areas for their personal use. Middle Eastern pharaohs regulated waterfowl hunting through licenses. Ashoka, India's first Buddhist king, set out edicts on pillars, which dotted the entire countryside. Among the Ashoka Pillar Edicts were rules that protected animals and advocated restraint to minimize overuse of natural resources. In Medieval Europe, kings and princes established royal forests for hunting—11,000 hectares in the eleventh century alone. Meanwhile, France's Forest Code (introduced in the 1300s), reserved timber for the use of the government, in this instance, the French Navy.

The beginning of the Industrial Revolution in the late eighteenth century, which brought the mechanization of production, spurred a wave of interest in the loss of nature and what it represented. William Wordsworth, one of the English romantic poets, deemed the Industrial Revolution an "outrage done to nature." Respect for nature was one of the recurring themes in the art and literature of the European Romantic Movement of the eighteenth and early nineteenth centuries. In 1836 Ralph Waldo Emerson published his first essay, "Nature," which inspired a parallel movement—Transcendentalism—among American authors and continued in the works of Henry David Thoreau, Margaret Fuller, Walt Whitman, and others.

The roots of modern western conservation efforts were born in the European colonial period. French naturalists Pierre Poivre, Philibert Commerson, and Bernardin Saint-Pierre were among the pioneers of the conservation movement in the mid-1700s. They witnessed the unprecedented scale of ecological change due to European expansion, and were concerned about the impact of deforestation on climate and species. The conservation measures they proposed were an amalgam of philosophies, drawing on Indian and Chinese forestry and horticulture practices, as well as European ones.

Alexander Von Humboldt, another early environmentalist, was a German explorer and geographer. Working in the late 1700s and early 1800s, he devoted his life to the conservation of the natural environment. Von Humboldt championed an ecological concept of relations between humans and the natural world drawn from Hindu philosophy. His work in South America highlighted the downstream consequences of cutting trees in upland areas, influencing governments to conserve upland forest reserves.

The conservation movement in the United States drew its roots from European ideals. A seminal event in the birth of the conservation movement was the public outcry in 1852 when the "Mother of the Forest," a giant sequoia tree, 300 feet high, 92 feet in circumference, and about 2,500 years old, was cut down for display in exhibitions and carnival sideshows. The tree grew in Calaveras Grove, California, an area that would eventually become part of Yosemite National Park. In 1864 George Perkins Marsh, sometimes considered America's first environmentalist, wrote *Man and Nature: The Earth as Modified by Human Action*, heralding forest preservation, and soil and water conservation.

Conservation efforts in the United States between 1870 and 1910 led to two major developments: the first was setting aside large tracts of forest as national parks, beginning with Yellowstone National Park in 1872; the second was the suppression of fire in forest management, a legacy that haunts U.S. forests to this day.

Two opposing viewpoints dominated American conservation pioneers. The "aesthetic" viewpoint—also known as the "Romantic Transcendental Conservation Ethic"—championed by the likes of John Muir, Ralph Waldo Emerson, and Henry David Thoreau, emphasized the importance of rare species, old growth wilderness areas, and the rights of wildlife (see Figure 5.1). Opposing this philosophy were the pragmatic views of Gifford Pinchot (who became the first head of the new U.S. Forest Service in 1898) and Theodore Roosevelt. These men forwarded the "Resource Conservation Ethic," a "multiple-use" concept for the nation's land and water, encouraging logging, grazing, wildlife and watershed protection, and recreation simultaneously. Their emphasis was on rapid productivity, abundance of game animals, and the right of access to resources.

These two philosophies were melded in part by Aldo Leopold in the middle of the twentieth century in a movement now called the "Evolutionary-Ecological Land Ethic." Leopold provided the

Figure 5.1 U.S. President Theodore Roosevelt and John Muir, founder of the Sierra Club, at Glacier Point, Yosemite National Park (*Underwood and Underwood, 1906*)

philosophical foundation for the development of conservation biology. In his writings, Leopold drew on the analogy of a watchmaker, noting that a watch is not a collection of independent parts, but a complex and integrated system of interdependent processes and components. Proper functioning of each part depends on the other components; together they make the watch function. Leopold stressed that a "wise tinkerer" saves all the parts, explaining that ecological processes are greater than the sum of individual species.

Modern U.S.-based conservation efforts combine parts of all three philosophies. Single individuals had a tremendous impact on the development of these efforts. Theodore Roosevelt, as Governor of New York State, fought to develop conservation strategies for the State's forests and rivers. Later, as President of the United States and working with his

Chief Forester, Gifford Pinchot, he brought conservation to the fore-front as a national priority for the first time. Rachel Louise Carson, a scientist and an eloquent writer, wrote *Silent Spring* in 1962, calling for an end to indiscriminate pesticide use and, more broadly, a change in the way we view nature. Her literary and scientific focus catalyzed new environmental laws and fostered the development of conservation biology, and also inspired public interest in the environment.

Growing concern about the environment eventually led to the United Nations Conference on the Human Environment, which was held in Stockholm in 1972. An outgrowth of the conference was the creation of the United Nations Environment Program (UNEP) as well as environment agencies as part of governments around the world. In the 1970s the United States passed four pivotal Acts, the Clean Water Act, the Clean Air Act, the Endangered Species Act, and the National Environmental Policy Act, which laid the foundation for environmental regulation. These acts served as models for legislation elsewhere.

PROTECT LAND

Protecting land and habitat is a central strategy for conserving biodiversity. There are over 108,000 protected areas around the world, covering 13.5 million square kilometers or about 11.5 percent of the planet, according to IUCN's World Commission on Protected Areas (see Figure 5.2). Preservation of the marine realm has lagged behind efforts to protect land. There are now 4,116 marine protected areas, but these protect less than 1 percent of the world's oceans.

At the Fourth World Parks Congress in Caracas, Venezuela, in 1992, countries were challenged to conserve at least 10 percent of their land. Many countries have since surpassed this target; however, the 10 percent protection level is somewhat deceptive. While more land is indeed protected than previously, the land that is protected still doesn't represent the diversity of biomes and ecosystems around the world. In the past, national parks were often created in places with spectacular scenery and little apparent economic value, such as mountains and deserts. Others were designed to manage game species rather than the habitats and ecosystem processes upon which these and other species ultimately depend. As a result, existing protected areas do not represent all ecosystems equally, and many have been nearly or completely lost due to conversion to agricultural and urban uses.

For example, in the United States, only 5.1 percent of land has been set aside as either a National Park, National Monument, or Wilderness Area (the highest level of protection according to IUCN categories),

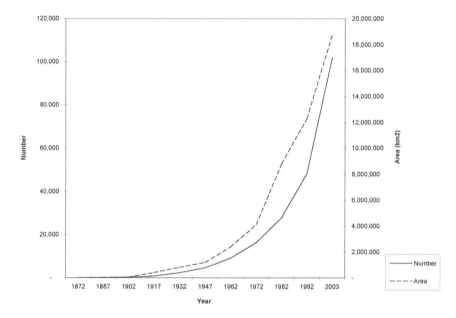

Figure 5.2 The number of protected areas in the world has grown dramatically since 1872 (*Laverty*) based on data from the WCPA

and most of these areas are found at higher elevations and on less productive soil (Dellasalla et al. 2001; Scott et al. 2001). Protection also varies widely from state to sate; in the eastern and central United States less than 1 percent is protected, while in Alaska about 35 percent of the land is protected. This trend is typical in many parts of the world, where the ecosystems most suitable for agriculture, such as dry tropical forests, prairies, and grasslands, are underrepresented in protected areas. Examining particular biomes, the amount of land protected within any one biome varies from 4.6 percent to 25.3 percent (Hoekstra et al. 2005). In tropical areas, lowland and montane, moist and wet forests were overrepresented, while dry forests were underrepresented (Green et al. 1997). Just over 5 percent of tropical rainforests are protected, while 29 of 63 countries in Asia, Africa, and Latin America have already lost 80 percent of their natural habitats (Soulé and Sanjayan 1998; Brooks et al. 2004).

───────────── ✍✍ ─────────────

WHAT IS A PROTECTED AREA?

A "protected area" as defined by the IUCN is "an area of land and/or sea especially dedicated to the protection and maintenance of biological diversity, and of natural and associated cultural resources, and managed through legal or other effective means."

IUCN has identified over 140 names for a "protected area"; however, these names often have different meanings in different countries. To provide greater international consistency and comparability, IUCN created six broad categories to reflect the level of protection and management of protected areas:

Category I: (a) Strict Nature Reserves, and (b) Wildlife Refuges, which are managed mainly for science and wilderness protection.

Category II: National Parks managed mainly for ecosystem protection and recreation.

Category III: Natural Monuments managed mainly for conservation through management intervention.

Category IV: Habitat or Species Management Areas are meant for active management to maintain a species or habitat.

Category V: Protected Landscapes or Seascapes are managed mainly for conservation and recreation. And

Category VI: Managed Resource Protected Areas are managed mainly for sustainable use of natural resources.

—————————————————————— ✒ ——————————————————————

Size Matters: Are Protected Areas Too Small?

The issue of size of protected areas is closely linked to concerns for identifying and protecting critical habitats. Most existing protected areas do not represent the full range of biomes on the planet, and most are too small to sustain the biodiversity found within their boundaries. According to IUCN's World Commission of Protected Areas, of the 108,000 protected areas in the world, 59 percent of them are very small, each covering an area less than 100 hectares (or 247.1 acres), less than a third the size of Manhattan's Central Park. As a general rule, larger species, especially carnivores with high metabolic demands (e.g., cougars, grizzly bears, and harpy eagles), need large areas to sustain the prey populations upon which they depend. Large areas also include and sustain greater diversity and redundancy of habitats, providing for a greater amount of biodiversity and the resiliency of more natural ecosystem processes. Consequently, most conservationists conclude that large contiguous areas of protected habitats are a critical tool for maintaining both species and ecosystem functions.

In the United States and Canada, 97 percent of protected areas are less than 10,000 hectares, and only 0.64 percent is larger than 100,000 hectares, a size considered big enough to sustain big carnivores and recover from large-scale disturbance.

Table 5.1 World's Largest Protected Areas

1. Northeast Greenland National Park, Greenland: 97.2 million hectares
2. Ar-Rub'al-Khali Wildlife Management Area, Saudi Arabia: 64 million hectares
3. Great Barrier Reef Marine Park, Australia: 34.5 million hectares
4. Northwestern Hawaiian Islands' Coral Reef Ecosystem Reserve, United States: Over 34 million hectares
5. Amazon Forest Reserve, Colombia: More than 32 million hectares
6. Qiang Tang Nature Reserve, China: Almost 25 million hectares
7. Cape Churchill Wildlife Management Area, Canada: Just under 14 million hectares
8. Northern Wildlife Management Zone, Saudi Arabia: 10 million hectares
9. Alto Orinoco-Casiquiare Biosphere Reserve, Venezuela and Bolivia: Over 8 million hectares
10. Vale do Javari Indigenous Area, Brazil: Just over 8 million hectares.

Source: IUCN-WCPA World Database of Protected Areas.

Not surprisingly the world's two largest protected areas, Northeast Greenland National Park at 97.2 million hectares, and Ar-Rub'al-Khali wildlife management area in Saudi Arabia (64 million hectares), protect Arctic tundra and Arabian desert respectively, hardly places desirable to people (see Table 5.1).

Protecting Individual Species

Since existing protected areas fail to include many species, conservation efforts in the 1970s and 1980s focused heavily on greater habitat protection for threatened populations and species. These projects aimed to identify and protect the minimum number of habitat areas that a population or species needs to avoid extinction, a goal shared by current "Habitat Conservation Plans" under the U.S. Endangered Species Act. Such conservation efforts have tended to focus on larger, more "charismatic" vertebrate species, such as spotted owls, gorillas, tigers, and giant pandas. Many conservationists argue that by protecting the habitats of these "flagship" or "umbrella" species, habitats for many other species are protected as well. Since the 1990s, however, greater emphasis has been placed on the direct analysis and conservation of all biodiversity. These efforts have been conducted in a large number of ways with no consensus yet on the best ways to identify and conserve biodiversity.

The first ever "global gap analysis" was presented at the Fifth World Parks Congress in Durban, South Africa, in 2003, to determine where the "gaps" in protection lie, or which species' range fell outside protected areas. The study examined the distribution of terrestrial vertebrates on the planet with respect to protected areas and found that the range of 1,400 of them lay outside of protected areas.

_____ ∽∂∾ _____

SAVED FROM EXTINCTION? REINTRODUCTIONS

Native to the deserts of the Arabian and Sinai Peninsula, the Arabian Oryx (*Oryx leucoryx*) is the smallest of the oryxes. This medium-sized antelope has slender horns that extend 2 to 3 feet in length, and a white coat with black-brown facial markings and legs. Prized for its delicate, long horns and coats, the Arabian Oryx was hunted with automatic weapons in the 1940s and became extinct in the wild by 1972.

Fortunately Arizona's Phoenix Zoo established a breeding herd in the 1960s and with these nine animals has helped to reintroduce the Arabian Oryx to its former range. Success has been mixed. While populations in Saudi Arabia and Israel have expanded, in Oman they have decreased dramatically. In 2007, UNESCO and IUCN removed Oman's Arabian Oryx Sanctuary from the World Heritage List due to the many threats facing the protected area.

While captive breeding programs at zoos, like that for the Arabian Oryx, have helped many species in crisis, this is an extremely costly way to protect biodiversity. The California Condor (*Gymnogyps californianus*) recovery program is considered one of the most successful. In 1982 there were 22 California condors remaining in the wild—these were eventually captured in 1987 to help rebuild the population. Reintroductions started in California in 1992 and in Arizona in 1996. The first pair bred in the wild in 2002. Currently, there is a captive population of 144 and a wild population of 141 birds. Captive breeding followed by reintroduction of the California Condor has cost around $40 million over a 20-year period.

_____ ∽∂∾ _____

Zoos, Botanical Gardens, Genebanks, and Seedbanks

Zoos, botanical gardens, genebanks, and seedbanks all offer ways to conserve biodiversity and have helped restore populations of near-extinct species; however, these reintroduction efforts are costly compared to conserving existing wild areas and a limited option for long-term species conservation. It also raises a philosophical issue, which is, if a species only survives in a zoo, have we really saved the species?

Gene banking is one mechanism to increase diversity in captive-bred populations, or to revive critically endangered species. Animal cells (including DNA, tissue samples, eggs, and sperm) are frozen at $-350F$ and stored in liquid nitrogen. Cells can later be thawed and still be functional, lasting hundreds of years or even longer. The San Diego Zoo began creating its "Frozen Zoo" in 1976. Their collection includes more than 350 species. The American Museum of Natural History launched the Ambrose Monell Tissue Collection in 2001. Eventually it will house over 1 million frozen tissue samples. The collection provides a key resource to support genetic research.

Seed banks are the parallel of gene banks for plants. They are an important resource both to supply genetic diversity to agricultural crops, and also to sustain endangered or even extinct plants. The viability of seeds ranges from decades to hundreds of years. The Millennium Seed Bank Project at the Royal Botanic Garden, Kew (United Kingdom), will eventually collect 10 percent of the world's plants.

CLONING ENDANGERED SPECIES

Since "Dolly the sheep" was cloned in 1996, people have wondered if it might be possible to clone endangered or even extinct species. The technique used to clone "Dolly" is called somatic cell nuclear transfer. A somatic cell is any adult cell other than an egg or sperm. To create the clone, scientists take an unfertilized egg and then transfer the nucleus or an entire somatic cell from the animal to be cloned into the egg. The fertilized egg is then put into a surrogate mother. Because of the need for a suitable surrogate mother, cloning efforts have focused on endangered species compatible with domestic animals, such as cows and sheep. Among the endangered species, scientists have cloned to date are a gaur, a banteng (Asian ox), and the European mouflon (an endangered sheep). Cloning, however, is a costly venture with a low success rate. For example, when the San Diego Zoo along with the Massachusetts biotech firm Advanced Cell Technologies, cloned a banteng, they initially implanted embryos in 30 cows. Sixteen cows became pregnant, but only two produced offspring and only one of those survived. Cloning also suffers from the same issues raised in captive breeding programs: how do you sustain genetic diversity and create a population that is self-sustaining, and what if their habitat has disappeared, have we really saved the species if it only survives in a zoo?

PROMOTE ENVIRONMENTAL EDUCATION AND STEWARDSHIP

Learning about biodiversity is the first step toward its protection. The more that people are aware of biodiversity and appreciate its value, the more likely it will be considered in people's decisions (see Figure 5.3). Connecting people to their environment through citizen's monitoring programs and helping people find ways to conserve biodiversity on their own land are just a few of the ways we can build a new appreciation of biodiversity.

Figure 5.3 Learning about biodiversity is the first step towards its protection. The more that people are aware of biodiversity and appreciate its value the more likely it will be considered in people's decisions (*Domroese©CBC-AMNH*. Used by permission)

Conservation on Private Lands

Animals don't recognize the boundaries between land that is inside and outside protected areas, so in many areas of the world, how land is managed outside protected areas is also critical for conservation, as a large percentage of land is privately owned. In countries like the United States, where government involvement in private land management is limited, conservation efforts must involve local landowners to effectively protect species and their habitat. Finding incentives to involve landowners in planning and implementing conservation strategies is crucial. One option is tax incentives for landowners who donate their land for conservation or create conservation easements. These incentives seem to be working as private land set aside for conservation is growing; areas preserved by local, state, and national land trusts increased by 54 percent to 37 million acres between 2000 and 2005 (Aldrich and Wyerman 2006).

Conservation Easements

One method to maintain open space and preserve farmland or natural areas is through the creation of easements. By establishing a conservation easement, landowners retain their land but restrict how it can be used. Typically, the land is held in trust either by a nonprofit organization or a state agency on behalf of the donor. The "land trust" is responsible for enforcing the restrictions on the land. Land trusts are nonprofits that help establish conservation easements on private lands and then act as land stewards ensuring the easements are maintained. The creation of conservation easements has been spearheaded by non-profit organizations, like Nature Conservancy, the Land Conservancy, and the Land Trust Alliance. Many smaller, local organizations as well as state land trusts have also been pivotal in increasing the land protected by conservation easements. In the United States, the number of local and state land trusts nearly quadrupled between 1982 and 2007, growing from 450 to 1600. In upstate New York, the state government protected 51,000 acres of forest in St. Lawrence County with an easement. Rayonier LLC, a forestry company, will continue sustainable logging on some parts of the land, while promoting recreation and protection of other parts. Since 1995 New York State has preserved over 1 million acres. The largest conservation easement to date in the United States protects 763,000 acres of Maine's Pingree forest and is overseen by the New England Forestry Foundation.

MALPAI BORDERLANDS GROUP

The Malpai Borderlands Group (MBG) (a collection of landowners, scientists, and other stakeholders dedicated to maintaining the health of a million-acre region in southern Arizona and New Mexico) is among those leading the way in the creation of conservation easements (see Figure 5.4). Formed in 1994 by a community of concerned ranchers, the group works to conserve ecological diversity through conservation easements and a variety of smaller projects while maintaining and promoting productivity of the land itself.

Years ago ranchers noticed the effects that development from nearby towns had on the landscape, and how the ecological consequences of this development impacted their livelihoods. The community had witnessed over a short span of time the transformation from mostly grasslands into mostly scrublands due to fragmentation and a lack of fires to limit the growth of woody species. So far, the group has obtained twelve easements totaling 75,000 acres from neighboring landowners, with the goal of expanding that area to 1 million acres. Four of these transactions were negotiated based on a concept invented by the MBG and referred to as "grassbanking." Instead of paying cash to a landowner for a tract of land, the group offers ranchers a land-use easement—in times of extreme drought, a rancher can graze his herd in less affected pastures on the Gray Ranch (comanaged by the MBG

Figure 5.4 The Malpai Borderlands Group (MBG)—a collection of landowners, scientists, and other stakeholders dedicated to maintaining the health of a million-acre region in southern Arizona and New Mexico—are among those leading the way in the creation of conservation easements (*Koy©CBC-AMNH.* Used by permission)

and the Animas Foundation). This protects the rancher's land in two ways: his pastureland can recover while his herd grazes elsewhere, and his land is off-limits to subdivision and development.

In addition to easements, the Malpai Borderlands Group oversees a variety of land management projects to restore grasslands and protect biodiversity. They set carefully controlled fires on areas where restoration is still possible, and their Watershed Improvement Project uses nothing but rocks and branches to build structures that redirect erosion and reduce water runoff. There is also a program underway to remove massive amounts of litter—as much as 80 backpacks in one day—left behind by undocumented aliens crossing into the United States over the Mexico border.

When the group initially formed, the protection of endangered species was not originally on the agenda. Today these animals figure prominently in planning for the region and their presence has proven beneficial in achieving the ranchers' landscape goals. (See also: http://www.malpaiborderlandsgroup.org.)

There are many conservation incentives designed to encourage conservation on private lands (see Table 5.2) One example is the development of "Safe Harbor" agreements between governments and landowners that promote conservation on private lands (often voluntary management to protect endangered species), while assuring landowners' future options for developing their land. The "Safe Harbor" concept was developed jointly by the Environmental Defense and the U.S. Fish and Wildlife Service. The first agreement was established in North Carolina in 1995 to protect the red-cockaded woodpecker (*Picoides borealis*). Similar programs also exist internationally. In Costa Rica landowners are paid for forest conservation and sustainable logging. In Mexico, Ecuador, and Columbia revenue from water charges are used to pay for forest conservation in critical watersheds.

MONITOR BIODIVERSITY

To conserve biodiversity effectively, we need to know where it is and understand how populations change over time. Because many species are in crisis, without effective monitoring it is hard to be able to stop a species loss before it is too late. In some cases we may not even know a species exists before it disappears. The rapid increase in many threats to biodiversity (such as ecosystem loss and fragmentation, invasive species unsustainable use, pollution, or climate change) necessitates a system that measures where biodiversity is and how it is faring. Strategies to conserve biodiversity ideally should be linked to the main threats. While hunting might not cause an entire forest to disappear, one or multiple species could vanish if that is the main threat they face. This may lead to what is called the "Empty Forest Syndrome"—while the forest still exists it is silent because all the species have been hunted. As Noss (1990) points out, the hierarchical nature of biodiversity dictates that we monitor biodiversity at many levels: from mapping the way that animal and plant communities are distributed across landscapes to identifying the composition of those communities, tracking the increase or decrease of species' populations, and measuring the genetic structure of those populations. Designing such a comprehensive monitoring protocol continues to be a challenge. However, only with such broadly gathered information can we design conservation strategies that reflect the most current threats to biodiversity.

CITIZEN'S MONITORING

Thanks to the Internet, citizen's monitoring of biodiversity has been growing by leaps and bounds. People can monitor everything from frogs to dragonflies to sea turtles; or even entire watersheds, rivers, or lakes. By

Table 5.2 Conservation Incentives for Private Landowners

- **Bay Farms Project**—Helps farmers make better use of nutrients in the soil, which is beneficial to the environment and to their bottom line.
- **Candidate Conservation Agreements with Assurances**—Landowners receive assurances that regulations won't change by committing to land management practices that protect at-risk species.
- **Carbon Sequestration**—Landowners receive money or carbon credits by storing carbon through landscaping choices (e.g., reforestation, cover crops, etc.).
- **Conservation Banking**—A landowner who protects habitat for an endangered species receives "credits" that can be sold to another landowner looking to develop an area where that species resides.
- **Conservation Reserve Enhancement Program**— Ranchers and farmers receive financial incentives or other assistance to restore habitat or retire their land.
- **Conservation Reserve Program**—Ranchers and farmers receive rent or other assistance to protect or restore sensitive land for a 10–15 year period.
- **Conservation Security Program**—Ranchers and farmers are given financial incentives or technical assistance for conservation practices on their land.
- **Environmental Quality Incentives Program**—Financial, technical, or educational assistance to landowners who institute environmentally sound practices on their land, such as better waste management.
- **Farm and Ranch Lands Protection Program**—Pay farmers and ranchers development rights to keep the land being used for agriculture.
- **Forest Land Enhancement Program**—Offers cost share assistance to private forest owners for conservation practices, such as planting trees, improving water quality, and controlling invasive species.
- **Grassland Reserve Program**—Rental or easement payment to landowners who commit to grassland restoration on their property or to keep grassland for grazing rather than developing it.
- **Healthy Forest Reserve Program**—Financial or technical assistance for private landowners who protect endangered species and restore habitat and generally improve forests on their land.
- **Lake Erie Special EQIP Project (EQSP)**—Farmers who want to develop a detailed conservation plan for their land can apply for technical assistance and per acre monetary grants.
- **Landowner Incentive Program**—Cost share assistance and professional advice for people or nonprofits who implement land management practices that restore habitat for species.
- **Private Stewardship Grants Program**—U.S. Fish and Wildlife Service works with landowners to protect habitat for all levels of at risk species. Technical help and cost sharing grants.
- **Safe Harbor Agreements**—Agreements between governments and landowners that promote conservation on private lands. They usually entail voluntary management to protect endangered species, while assuring landowners' future options for developing their land.

(*Continued*)

Table 5.2 (*Continued*)

- **Technical Service Providers**—This is for landowners already involved with one of the USDA's Natural Resources Conservation Service programs. Landowners can apply for technical assistance to adopt conservation measures on their land.
- **Wetlands Reserve Program**—Offers financial assistance to landowners with wetlands on or adjacent to their property. Pays landowners for long-term and permanent easements on agricultural land that houses wetlands or costs to restore wetlands in some cases.
- **Wildlife Habitat Incentives Program**—Cost share and technical assistance for landowners to protect or create wildlife habitat.

Source: Center for Conservation Incentives, Environmental Defense.

tapping into the energy of citizens we can monitor biodiversity over a much larger area than would be possible otherwise. Monitoring also gets people interested in biodiversity and connected to their environment directly.

Audubon Christmas Bird Count: The oldest citizen-monitoring program in the United States and probably the world. People have been conducting a bird count from mid-December to early January for the last 100 years. (http://www.audubon.org/bird/cbc/)

Frogwatch U.S.A.: This frog- and toad-monitoring program allows people around the United States to help scientists understand and conserve amphibians. (http://www.nwf.org/frogwatchUSA/)

The Ocean Conservancy's International Coastal Cleanup: The world's largest single-day volunteer effort aim's to clean up the world's coastlines. (http://www.oceanconservancy.org/icc)

Dragonfly Monitoring Network: This program monitors the distribution and abundance of dragonflies and damselflies in the Greater Chicago area. (http://www.anisoptera.org/)

Monterey Bay Citizen Watershed Monitoring: A network of 20 organizations that monitor the watersheds draining into the Monterey Bay National Marine Sanctuary. (http://montereybay.noaa.gov/monitoringnetwork/welcome.html)

Citizen Stream Monitoring: This program began in 1998 with just 17 volunteers and as of 2007 had grown to over 400 volunteers who monitor nearly 700 sites across Minnesota's ten major river basins. (http://www.pca.state.mn.us/water/csmp.html)

Affect Environmental Policy and Legislation

Laws, and the legislative process, are a mechanism for society to codify its principles, aspirations, and structures. From the standpoint of conservation, legally recognized protection of biodiversity at global, national, regional, and local levels is essential on both philosophical and practical levels. International agreements with the force of law, such as the Convention on International Trade in Endangered Species of Wild Fauna and Flora or CITES, simultaneously signal the world's commitment to end the trafficking of endangered wildlife and provide mechanisms for the convention's signatories to reach that common goal (see Table 5.3). CITES monitors over 30,000 species of plants and animals endangered by the wildlife trade. Countries that are a party to CITES agree to implement the Convention's licensing system to control the import and export of listed species. National laws, such as the U.S. Endangered Species Act, which articulates that species have "esthetic, ecological, educational, historical, recreational, and scientific value to the Nation and its people," are equally critical in protecting species and their habitats. Conservationists must continue to understand and involve themselves in the legislative process in order to protect existing environmental laws, encourage additional legal protections for biodiversity, and foster the inclusion of conservation ethics in law.

Tax Plastic Bags

Plastic bag litter is a huge problem around the world. Littered plastic bags are not only an eyesore but also spell bad news for wildlife. Plastic bags often find their way into oceans and rivers where birds and mammals mistake them for food. In 2002, the Irish government instituted the "plastax," requiring shoppers to pay 15 cents for a plastic bag (as of 2008 the tax has increased to 33 cents per bag.) In just a few months, plastic bag use decreased by 90 percent and the tax generated a whopping 3.5 million Euros. While the tax was created to change consumer behavior, the substantial revenue generated was an unexpected bonus, which now goes into an Environmental Fund. The tax has eliminated the use of 1.2 billion plastic bags a year, or 18 million liters of oil—since oil is used to produce plastic. It has also significantly reduced the plastic waste in landfills. In 2008 Ireland was considering taxes on bank machine receipts and chewing gum.

Some Fast Facts About Plastic Bags

- 500 billion bags are consumed worldwide each year.
- Only about 3 percent of plastic bags are recycled, the rest end up in landfills where it takes 500 years for a bag to decay.
- The United States consumes 100 billion plastic shopping bags annually. An estimated 12 million barrels of oil is required to make them. The

Table 5.3 Major International Treaties That Protect Biodiversity

Year	International Agreement	Treaty Focus	Number of Contracting Parties (as of May 2008)
Adopted 1972, entered in force 1975	UNESCO World Heritage Convention	Identify, protect, and preserve sites of cultural and natural heritage; 851 sites are listed: 661 cultural, 166 natural, and 25 mixed.	185
Adopted 1971, entered into force 1975	Convention on Wetlands of International Importance (Ramsar Convention)	To conserve and promote the wise use of wetlands and their resources; 1743 wetland sites covering 161 million hectares.	158
Adopted 1992, entered into force 1994	Convention on Biological Diversity (CBD)	To sustain the rich diversity of life on earth.	190
Adopted 2000, entered into force 2003	Cartagena Protocol on Biosafety, part of the CBD	To protect biodiversity from organisms created by biotechnology.	147
Adopted in 1973, entered into force 1975	Convention on International Trade in Endangered Species (CITES)	To ensure international trade in wild plants and animals does not threaten their survival	172
Adopted 1946	International Convention for the Regulation of Whaling (ICRW)	To conserve whale stocks and make possible the development of the whaling industry.	79
Adopted 1992, entered into force 1994	United Nations Convention on Climate Change	To share information about Greenhouse gases (GHG); Encourages countries to launch strategies to reduce GHG emissions.	192

Table 5.3 (*Continued*)

Adopted 1997, entered into force 2005	The Kyoto Protocol of the Convention on Climate Change	Commits countries to reduce GHG emissions.	84
Adopted 1982, entered into force 1994.	United Nations Convention on the Law of the Sea (UNCLOS)	Guidelines for the use and management of the world's oceans and marine resources.	155
Adopted 1987, entered into force 1989.	Montreal Protocol on Substances That Deplete the Ozone Layer	Controls the production of ozone-depleting substances. There have been a series of amendments to the Protocol in: London, Copenhagen, Montreal, and Beijing	191
Adopted 1973, modified by Protocol in 1978	International Convention for the Prevention of Pollution from Ships (MARPOL 73/78); 6 annexes	To prevent pollution from ships, such as oil, chemicals, and sewage.	Ann. I/II:146 Ann III:128 Ann IV:118 Ann V:134 Ann VI: 49
Adopted 1994, entered into force in 1996	United Nations Convention to Combat Desertification	To reverse and prevent desertification and to mitigate the effects of drought.	190

Source: Compiled by Laverty.

average family accumulates 60 plastic bags in only four trips to the grocery store.

- In a dramatic move to stem a tide of 60,000 metric tons of plastic bag and plastic utensil waste per year, Taiwan banned both in 2003.
- After Bangladesh discovered that plastic bags were clogging the country's main drainage system and causing severe floods in 1988 and 1998, they also banned plastic bags.
- Windblown plastic bags are so prevalent in Africa that a cottage industry has sprung up harvesting bags and using them to weave hats and bags.

- Places with total bans on plastic bags include Singapore, Bangladesh, South Africa, Tanzania, and most recently China, in preparation for the 2008 Olympics. Many other countries have partial bans or other incentives to reduce the use of plastic bags, including Australia and Italy. Other countries considering banning plastic bags are Israel, Canada, Botswana, and Kenya.

- San Francisco banned nonrecyclable plastic bags at large supermarkets and pharmacies in 2007, but the proposed legislation to tax plastic bags failed. In New York large stores that give out plastic bags must also collect and recycle them as of 2008. In Paris nonbiodegradable bags were banned in 2007. In June 2007, Modbury became the first town in England to ban plastic bags; since then about 80 towns have followed their lead, and even London is expected to follow suit.

- A growing trend is to ship plastic and other waste to less developed countries, like India and China. Rather than being recycled they are cheaply incinerated under more lax environmental laws.

--- ᶜᵒᵛ ---

SWITCH LIGHT BULBS

Australia will be the first place to entirely ban incandescent light bulbs in a move to reduce energy consumption. Brazil had already established incentives to encourage the use of compact fluorescent lamps (CFL) rather than incandescent lamps. Traditional bulbs, modeled after Thomas Edison's design of the nineteenth century, are inefficient because they waste a large amount of energy as heat. Electricity causes a filament in the bulb to heat up and glow, producing light. In contrast, CFL bulbs don't produce heat and so consume a quarter of the energy of their incandescent counterparts and last ten times longer. This type of bulb is filled with a gas that produces UV light upon interacting with electricity. In North America, look for bulbs labeled "EnergyStar." Once all bulbs have been replaced with newer, more efficient ones, the energy savings are expected to be significant. Lower energy needs translates into lower carbon emissions from power plants. Other places phasing out incandescent bulbs include: Canada, Ireland, New Zealand, Russia, and the United States.

LED bulbs are also revolutionizing the way we light up the night. LED is short for light-emitting diode. Unlike incandescent bulbs, these tiny bulbs are made out of semiconducting material that creates light through the movement of electrons. They last longer than incandescent bulbs whose filament burns out; they also generate little heat. A city that replaces old bulbs in 100 traffic signals with LED lights will use 93 percent less energy. Semiconductors were once too expensive for common uses but now prices have plummeted and LED lights are finding their way into everyday products, such as: traffic lights, holiday lights, bike lights, flashlights, signs, clocks, cell phones, and computers. They can have a very long life span upward of

100,000 to a 1,000,000 hours compared with 10,000 hours for fluorescent tubes and 1,000 to 2,000 hours for incandescent bulbs. They also can cycle on and off without reducing their lifespan substantially and are not as easily damaged. Though somewhat more expensive initially, LED lights easily outperform their incandescent counterparts due to their longer lifespan.

_____ ⌒⌒ _____

STABILIZE HUMAN POPULATIONS

From 1.6 billion people in 1900, the world population grew more than threefold, topping 6 billion at the end of the twentieth century. Having doubled since 1960, the world's population is expected to reach 9 billion people by 2050. If any sustainable balance between the world's biodiversity, ecological systems, and humans is to be found, we must stabilize human population growth. While consumption of the world's resources is still dominated by people living in rich, developed countries, the needs of the growing populations in less developed nations are nevertheless taking an enormous toll on the global environment. Some actions to help stabilize populations, particularly in the fastest growing nations, include improving the education and status of women; increasing the survival and health of children; and providing easy access to family planning resources.

CONSUME LESS

Any discussion of solving the problem of a swelling global human population must be accompanied by a parallel discussion of the patterns of human consumption of Earth's resources. Heavy consumption of resources is deeply engrained in people's lifestyles in the United States, as well as much of the developed world. Often they are almost imperceptible to us, but cumulatively the habits of daily life wreak havoc on the natural world. According to the Sierra Club's Sprawl Index, urban sprawl destroys 1 million acres of open space in the United States each year, fragmenting wildlife habitat and isolating populations of species. This dispersed pattern of settlement forces residents to use more energy to transport themselves between the home and all other destinations, while also increasing the cost of government services, such as garbage pickup and mail delivery, both of which contribute to reduced air quality and global climate change.

Increasingly, people are feeling the need to disconnect themselves from the senseless pursuit of wealth and from these cycles of consumption. The movement, known as *voluntary simplicity*, centers on simplifying one's life by examining what is actually necessary to live and eliminating things that are superfluous. Choices vary from person to person; they include shopping less, reducing your debt, downsizing your home, and reducing your work hours. Many find the changes are both psychologically and financially very rewarding. Voluntary simplicity is marked by intentional frugality and deliberate choices, in contrast to others who are just trying to make ends meet. Neither should it be confused with self-deprivation, but rather learning to be content with what is enough and shunning the excesses of modern life. In choosing to consume less, practitioners of voluntary simplicity have a smaller *ecological footprint* as they use less resources and require less energy.

There are various groups that have surfaced in recent years that espouse the ideas of voluntary simplicity. The Center for a New American Dream, based in Maryland, works with individuals and businesses to change the face of consumer culture, empower people by getting them more involved in the political process and their communities, and raise consciousness toward achieving more sustainable ways of living. Their Web site contains a wealth of information on decreasing junk mail, environmentally preferable purchasing, and alternative gift giving for the holidays.

A group of individuals in San Francisco called "the Compact" go beyond recycling and "green shopping" to take a stand against rampant consumerism. The Compact made a pact in 2006 to not buy anything new for the whole year, and it appears as though many of the members remain committed to the idea. The group has meetings and an on-line community where they offer support to help each other meet the common goals of waste reduction, decluttering, supporting the local economy, and, generally, simplifying their lives. "Compacters" are permitted to borrow or buy what they need as long as the item is secondhand. There are some exceptions—food, medicine, cleaning supplies—and other technical issues as they arise that are discussed and decided upon by the whole group. One concept that participants have discovered through this experiment is the degree to which shopping is a psychological need more than a way to meet physical needs. As the idea of the Compact catches on, more and more regional groups have sprung up all over the world.

Of course, consumption also creates tremendous quantities of waste. According to UNEP's Global Environment Outlook, in the early 1990s, the annual global output of hazardous wastes from chemical production, energy production, pulp and paper factories, mining, and leather and tanning production was about 400 million tons, with about three-quarters of that coming from the industrialized nations. If we are to lessen the impact that each one of us has on the Earth each day, we must simultaneously reduce the amount of goods and services that we consume daily and develop new, appropriate technologies that create goods and services at a smaller cost to the global environment.

───────────────── ∽∂∾ ─────────────────

HUMAN ECOLOGICAL FOOTPRINT

The "ecological footprint" model, developed by Mathis Wackernagel and William Rees, measures the demand people make on natural resources. It can be calculated for an individual, or all the individuals in a city, region, or country. Footprints vary widely among nations. It can be expressed in numbers of planets, where one planet equals the total biologically productive capacity of the Earth in one year. In other words, the number of planets needed to support humanity if everyone consumed natural resources at the rate of the group in question.

Today the human footprint is 23 percent larger than what the planet generates. The global ecological footprint in 2003 was 1.25 planets; that is, we are already using more resources than the planet is capable of sustaining. This means that humanity as a whole consumed resources at a level equal to the capacity of 13.9 billion global hectares, or 2.2 hectares per person. Since Earth's biologically productive area, or biocapacity, is 11.3 billion global hectares, providing an average of 1.8 global hectares per person, the global footprint measurement shows that humanity is exceeding the planet's biological capacity by 0.5 global hectares per person; in other words, we are using natural resources faster than they are regenerated. The global ecological footprint increased about 160 percent from 1961 to 2003, faster than the rate of population growth. Of course, while the global footprint measures the impact of the average person, in reality there is enormous disparity in the distribution of consumption. With people in some countries consuming few resources, while those in others consuming excess resources. (See also, http://www.footprintnetwork.org/gfn_sub.php?content=global_footprint)

───────────────── ∽∂∾ ─────────────────

Think About What You Eat

We are fortunate that nowadays we have an abundance of foods available to us throughout the year. Indeed, seasons and regions no longer limit our choices, at least in the developed world, as a glance at grocery displays and restaurant menus shows. But how does the availability of once-seasonal produce and foods from the farthest reaches of the planet impact our experience of foods, their nutritional value, and the environment? More and more, people are beginning to ask where their food comes from, and recognizing how it affects their well-being and the well-being of their community, and its vital link to nature and biodiversity.

In the last half century, the variety of grains, vegetables, and fruits grown in the United States has dwindled to a narrow selection of foods best suited for commercial production and marketing—those with higher yields, long-distance durability, longer shelf life, and standardized size and color. This transition has resulted in an estimated loss of approximately 75 percent of the genetic diversity among agricultural crops. The disappearance of diverse cultivars has meant not just a loss of unique flavors and textures, but also a precipitous drop in genetic variability. Species that encompass a wide range of genetic traits are better able to withstand threats posed by disease, pests, and environmental conditions, and our increased reliance on a handful of crops grown in uniform monocultures (one crop that is grown over large areas) means we are also reliant on a dangerously limited agricultural gene pool.

This loss of diversity ripples far beyond the homogeneity of farm fields. It also affects the variety of animals able to live on and around farmland, including birds, insects, small mammals, and others that provide invaluable services. For example, scientists estimate that pollination by insects, especially bees, is worth billions to the U.S. economy each year. Soil insects and other organisms enrich and fertilize the land. Areas lacking these natural populations are often reliant on massive amounts of chemical pesticides and fertilizers. The good news is that increased interest in more sustainable agriculture methods has led a growing number of farms to incorporate a greater variety of crops, use few or no chemicals, and conserve natural habitat, thus supporting healthy populations of pollinators, predator species that act as natural pesticides, and other benefits from biodiversity.

By seeking out seasonally available and locally grown foods, asking informed questions about farming and fishing methods, and considering the resources involved in bringing the foods we buy to market, we help restore and retain healthy ecosystems and strengthen local economies. We have a wealth of opportunities to meet growers and local food distributors—through farmers' markets, community-supported agriculture programs (CSAs), food co-ops, farm stands, and pick-your-own orchards. Get to know the people who produce and sell the foods you eat. In grocery stores, too, it's important to ask questions and look for locally produced foods. In the process of seeking out the bounty of foods available from local sources, we reap the benefits of a more diverse diet full of fresh foods, intense flavors, and healthful choices. Whether it's a native fruit, such as beach plum, a New York-bred heritage turkey, an heirloom variety of fruit or vegetable, or a different type of grain, your food choices can promote healthier people and a healthier planet.

The way that we eat reflects the increasing amounts of resources we use in our daily lives, as well as the large disparities in the resources it takes to feed a single person in different countries. For example, the Audubon Society recently reported that the Earth could feed 10 billion people eating as the citizens of India do, 5 billion as the Italians do, but just 2.5 billion eating as the Americans do. Nowhere is this more apparent than in meat consumption. As economies grow and people become more affluent, meat consumption typically increases. In 1900, 10 percent of the world's grain went to feed animals. By the 1990s that proportion had risen to 45 percent (Riebel and Jacobsen 2002). As we transition to meat-heavy diets, it takes almost four times more calories to feed each person. Some estimate that it can take as much as 2,500 gallons of water to produce one pound of beef, particularly if harvested from cattle raised in California, Arizona, or Colorado. Most of this water is used to irrigate grainfields and grass pastures grown in areas that are otherwise arid. Rather than consume local produce, Americans commonly eat food that is transported huge distances before it arrives on our tables—a hamburger served in Seattle, Washington, contains meat from Texas and Colorado, lettuce and tomatoes from California, wheat from Idaho, corn from Nebraska, and salt from Louisiana (Riebel and Jacobsen 2002).

Include Biodiversity in Economic Measures

> What is the logic of extracting diminishing resources in order to create capital to finance more consumption and demand on those same diminishing resources? (Paul Hawken, 1993, 5)

Change our Measures of Progress

Classic economic theory measures economic progress with continuous gains: increased production, consumption, and profits. Many of the most common economic measures ignore the decrease or loss of natural resources. The most basic and widely used of those indicators, gross national product (GNP), is no exception. As natural resources are consumed and the environments' ability to support healthy ecosystems is reduced, our ability to keep using those resources is also diminished, but current methods of calculating GNP do not reflect this.

Include the Value of Biodiversity in Markets

The value of biodiversity and the cost of its loss are generally excluded from commercial markets. Though we know that biodiversity and the services it provides to society are crucial to human well-being, we often take them for granted and treat them as if they are free. By incorporating biodiversity into economic markets, cost-benefit analyses, and policy, it will reinforce biodiversity's value. An attempt to estimate the value of ecosystem services was made by Costanza and several other economists in 1997. According to the study, the earth provides a minimum of $16 to $54 trillion worth of "services" to humans per year, based on the value of 15 ecosystem services and two goods in 16 biomes. Some scientists and social scientists have disputed the quantitative conclusions of this study; among the criticisms of the study is the difficulty of "scaling up" across ecosystems and types of services. Nevertheless, the study by Costanza is an important first effort to estimate the global economic contribution of ecosystem services. Trees are worth more in a big city than in a rural area. The Neighborhood Tree Survey in New York City assessed 322 trees at a total value of $1,038,458, with the range per tree being from $54–$23,069. Assessment categories included the physical condition of the tree; properties such as age, height, diameter; pollution removal, and carbon storage.

Products Should Reflect Their True Costs

Markets often fail to account for many of the costs associated with production and consumption. For example, when you drive your car and burn gasoline, who pays for the costs associated with the respiratory illnesses from the poor air quality that you are causing? What about the costs of a changing climate to which the CO_2 and NO_x coming from your tailpipe are contributing? These costs—considered externalities—are rarely included in the price paid by consumers. Alfred Pigou, an economist in the first half of the 1900s, formalized our thinking about these external costs; that is, costs that are left out of the modern economy. Following this logic, biodiversity and the environment are not protected because their value (or the cost of their loss) is not included in the pricing structures that shape business decisions and consumer behavior.

Abolish Perverse Subsidies and Incentives

Government policies sometimes unwittingly encourage or even pay for environmentally destructive practices. The U.S. Forest Service, for example, in fulfilling its mandate to provide the logging industry with access to the National Forests, winds up spending more on building roads than it recovers in logging concession fees—a windfall for the logging companies, which don't have to pay for their roads. In Brazil, a government policy designed to encourage settlement of the Amazonian frontier gave pioneers free land if they cleared more than half of the forest from their properties. It is estimated that the global sum of all of these destructive subsidies is $2 trillion USD annually (Myers & Kent 2001).

WHAT YOU CAN DO

All the bad news related to our misuse of the environment can seem overwhelming and the problems too complex, but there are solutions. People are making changes in their lives to conserve resources and protect biodiversity. Even small changes can make a difference. The following guidelines detail actions you can take to lessen your impact on the planet. Focus on the ones that you feel comfortable with, some changes are more difficult than others; most of us can't just trade in our cars for a new, fuel-efficient model, but we can choose to drive less or buy local food. Keep an open mind and remember that any improvement you make is 100 percent better than doing nothing at all (see Table 5.4).

Reduce/Reuse/Recycle

Reduce the amount of waste you create, reuse materials for other purposes, and recycle as much as possible. At the grocery store, avoid creating excess garbage by buying in bulk and avoiding items with excessive packaging. Don't forget to bring along your own bags; keep a supply of extra bags in your car if you drive to do your shopping. Take a portable mug with you for coffee instead of constantly accepting disposable cups. Some places even give a discount for your mug-toting efforts. If you regularly request a "doggy bag" when eating at restaurants, try to remember to bring your own container along for leftovers.

Start a compost pile in your backyard, or a vermicomposting bin using worms in your home, which is a more viable option for city dwellers.

Table 5.4 Ten Things You Can do to Conserve Biodiversity

1. Recycle, Reduce, Reuse
2. Eat Local, Seasonal, and Organic Foods
3. Travel Responsibly
4. Conserve Energy
5. Conserve Water
6. Reconsider the Daily Commute
7. Invest in Socially Responsible Companies
8. Consume Less
9. Green Up Your Home and Garden
10. Create Sustainable Communities.

Source: Compiled by the authors.

The Environmental Protection Agency (EPA) estimates that 25 percent of household waste generated in the United States is compostable, but most of it is currently land filled with the rest of our garbage. Some cities, such as Berkeley and California in the United States (and Toronto and Halifax in Canada), have begun to collect and compost organic waste.

Recycle items accepted by your local garbage collector or at a drop-off center. In most places, commonly accepted items include: plastic and glass bottles, newspapers, white paper, corrugated cardboard, aluminum cans, and steel. Many places will even reimburse you for scrap metal and aluminum, and various states honor refunds on beverage bottles.

Donate unwanted clothing, books, toys, and other household objects to thrift stores or similar organizations. Nonworking computers can be donated to organizations that repair electronics and later distribute them for use by underfinanced schools. There are a variety of programs for old cell phones, which are either dismantled for recycling or reprogrammed for use. Take old fixtures and construction materials to materials reuse warehouses. Join the Freecycle network—an on-line "gifting" network that matches unwanted items with people that can use them.

Invest in rechargeable batteries and a charger—your initial investment of $30 will save you over the long term. At the end of their useful life, bring them to your local electronics store, which typically accept all types of rechargeable batteries (e.g., cell phones, laptops, power tools, etc).

Reuse packing materials and Styrofoam peanuts or take them to a shipping company who will gladly accept them.

Go Paperless

Put an end to all the junk mail that piles up in your mailbox. Print double-sided, if your printer allows it, or onto the blank side of "used" paper. Better yet, don't print at all and keep documents in folders on your computer or on a CD. Use e-mail to send documents instead of fax machines, or subscribe to an electronic fax service. When printing from the Internet, use the Print Preview option to avoid printing out several pages that you don't need. Stop taking bank receipts at the Automatic Teller Machine (ATM). An estimated 375,000 receipts are thrown away each year in the United States, laid end-to-end they would circle the earth 15 times (Rogers and Kostigen 2007).

Recycling: Tackling the Misconceptions

There is considerable debate over whether or not recycling is more economical and environmental than land filling, with some people even questioning whether the process occurs at all. Most people agree that recycling is beneficial, but every so often, a study emerges proclaiming the opposite. Environmental Defense posits that these "antirecyclers" are people in the waste-hauling and incineration industries, consumer-packaging producers and various product manufacturers—and the policy think tanks they fund—who are concerned about how the competition of recycling could impact their livelihoods. Only a few large companies dominate the waste collection market, and these companies stand to lose the most if recycling becomes popular.

What Are the Recycling Critics Saying?

The biggest point of contention is: economics. Critics argue it costs more to recycle than to just cart disposables to a landfill. They also seriously underestimate the energy savings and pollution reduction that comes from recycling material instead of processing virgin materials. Some critics even contend that recycling does little to conserve resources, frequently citing trees and the fact that we plant more trees now than we cut down for paper production. Landfills themselves are often shown in a rosy light as being cheap and environmentally safe receptacles for garbage. The arguments are often one-sided, and the complete data paint a different picture.

Economically speaking, many feel that recycling should be a free service. Recycling like any government service has a price. Landfills and incinerators aren't free—these expenses are just less obvious. In some places in the United States, there is a direct charge to pick up recycling but not for garbage, and so people assume it's "free". In reality, they are either already paying for garbage disposal through their property taxes, or the local government has subsidized waste collection for that area, in effect making recycling appear as though it is more costly. An alternative waste collection program functions as "pay as you throw" systems, or PAYT. In PAYT, the individuals pay for the amount of garbage they produce, thus providing a direct incentive for residents to reduce their waste via recycling and/or composting. There are currently over 7,000 communities in the U.S. that use the PAYT system for garbage collection. Cities where there is limited landfill space like Toronto have adopted organic waste collection as well as recycling as a way to reduce the amount of waste that is sent to landfills.

Recycling Benefits

Recycling is also more effective in some areas than others; New York City has around a 20 percent diversion rate (that is the amount of waste that is diverted to recycling), while other cities have achieved a rate over 60 percent. The more products recycled, the more efficient the process and the lower the costs. The energy conserved by recycling versus that produced from incineration is staggering, particularly for aluminum. Recycling aluminum eliminates the need to strip-mine for metal ore, an extremely energy intensive and polluting process. After initial start-up costs, recycling creates more jobs than waste disposal and provides many long-term benefits. According to the U.S. Environmental Protection Agency (EPA) between 1990 and 2003 the amount diverted from landfills more than doubled, increasing from 34 to 72 million tons.

Remanufactured and Refurbished Goods

The idea behind remanufacturing is to take used goods, restore them to "like new" condition, and resell them. Refurbished electronics, like computers and cell phones, are becoming popular in the consumer market, but many different manufacturers have adopted this process. Examples include: *Interface*, which uses reclaimed materials to produce carpet tiles' vinyl backing; *Caterpillar's* diesel engine remanufacturing operation, one of the company's fastest-growing units, acquired $1 billion train car remanufacturer, Progress Rail. *Hanover Compressor* remanufactured its fleet of skids to meet demand. *Xerox* remanufactured 70,000 machines and diverted 128 million tons of material.

CONSUME WISELY

One way that individuals can exercise a little control over unsustainable manufacturing practices is through their buying power. Consider your purchasing habits in this light and put your money where your conscience lies. Big business has begun to sing a different tune in the new millennium in response to the shopping habits of smart consumers.

EAT LOCAL, SEASONAL, AND ORGANIC

Buy locally produced foods. You can enjoy fresh, seasonal produce while decreasing fossil fuel use for transport and supporting the local economy.

Buy organic foods. Organic agriculture practices avoid the use of chemical pesticides and fertilizers and emphasize natural techniques that maintain healthy ecosystems.

Grow your own food. Grow herbs on your windowsill or vegetables on your roof or in your yard.

Shop at a farmers' markets or join a co-op or CSA (community supported agriculture group). Farmers' markets are fun places to buy locally produced food and meet the people who produce it! Find farmers' markets, co-ops, CSAs, and more at www.localharvest.org.

Buy shade-grown coffee. Or better yet, buy organic, fair-trade, shade-grown coffee. Increasingly available in stores and coffee shops, shade-grown coffee is grown on farms that retain trees that provide crucial habitat for migratory birds.

Curb your carnivorous consumptions. Meat production requires large amounts of natural resources. For example, growing a pound of corn requires 100–250 gallons of water, while growing the grain to produce a pound of beef requires 2,000–8,500 gallons. In addition, livestock factory farms are a major source of air and water pollution and also erode topsoil.

Choose sustainable seafood. Many seafood stocks are depleted or harvested in environmentally harmful ways. Download a handy wallet card to help you make better seafood choices at http://www.environmentaldefense .org/seafood/fishhome.cfm, http://www.mbayaq.org/cr/seafoodwatch. asp, http://www.audubon.org/campaign/lo/seafood/index.html, and http://www.blueocean.org/Seafood.

Select natural beauty care products. When shopping for beauty care products and cosmetics, read the labels and avoid chemicals such as phthalates. See the Environmental Working Group's study for details.

Simplify holidays. Holidays and special events are particularly problematic for treading lightly on the earth. Avoid purchasing plastic cups for your parties, but if you must, have everyone write their name on it so they don't lose it. Buy plasticware made of sugarcane or corn, so it biodegrades. Compost your Christmas tree or dispose of it with yard waste if your locality has composting facilities. Be creative with gifts—does that plastic bobble-head Chihuahua really remind you of your loved one? Or maybe a donation in their name to an organization they care about would speak louder. Create "coupons" for chores you will help out with, or give gift certificates or tickets to museums. Opt for usable items, like a box of homemade cookies, or some quality organic olive oil soaps or soy candles. Why buy wrapping paper when you can recycle old magazines and calendars into some quite stunning gift wrap, or use reusable tins or pretty cloth bags—a gift within a gift. On Valentine's Day, skip the roses (which are often flown in from far-flung countries) and instead indulge in organic, fair-trade chocolates.

Request recycled diamonds and gold. Few people are aware of the extent to which their purchases of gold and diamond jewelry aid in the destruction of ecosystems. Forests are clear-cut to make way for mining operations, and waterways become contaminated with high levels of acid and cyanide, destroying essential habitats for wildlife on the land and in the lakes. They leave a long legacy of pollution; mines from Roman times are still producing acid. Seek Out Alternatives at http://www.ethicalmetalsmiths. com or http://www.greenkarat.com.

Buy recycled. The success of recycling is largely determined by markets; recycling depends on people buying recycled products. Purchase paper and paper products that are made of recycled material, select those with high levels of "post-consumer" recycled content. Shop for "recycled" or "used" goods and clothing by visiting thrift stores, flea markets, and garage sales.

Buy clothing made from organic or recycled fabrics, and choose things that last. As consumers support organic products, producers grow more organic produce and in turn the easier it is for manufacturers to find enough supply for their products.

HEALTHY EATING FOR YOU AND THE PLANET

Many countries have legislation that defines certified organic standards. In addition to buying *USDA-certified organic foods,* look for the *Certified Naturally Grown* label, which has comparable standards but is more affordable for small, independent farmers. In Canada, look for the *Canada Organic* label. There are also some international organic certification associations. Ultimately, the best way to know what is in your food is to talk to your farmer or grocer.

Avoiding Pesticides in Produce

If you're concerned about the high cost of organic food, here is a guide to indicate which fruits and vegetables typically have the highest pesticide levels. If you can only buy some organic produce, these are the ones to choose. Your consumer support can make a difference! As demand increases for sustainably grown foods, prices for organic food should fall.

The following fruits and vegetables consistently have *high* levels of pesticide residues—choose organic or other, low-pesticide options when possible.

Fruits: Apples, Cantaloupe, Cherries, Grapes, Nectarines, Peaches, Pears, Raspberries, and Strawberries.

Vegetables: Bell Peppers, Celery, Green Beans, Potatoes, Spinach, and Winter Squash.

The following fruits and vegetables consistently have *low* levels of pesticide residues.

Fruits: Bananas, Kiwi, Mangos, Papaya, and Pineapples.

Vegetables: Asparagus, Avocados, Broccoli, Cauliflower, Corn, Onions, and Peas (sweet).

Certification

Certification of products is a method to establish whether they have been produced in a sustainable way. From forests to clothes to tourism, labels from "organic" to "fair-trade" to "sustainable" indicate whether a product meets certain social, environmental, or ethical criteria. Certification is typically conducted by an independent third party according to a set of standards. Some industries have more than one certifying body, and thus it is important to consider how reputable the organization is conducting the certification. Also, certification can be costly, so don't rule out local growers who are operating organically but may not be able to afford to be certified. Other challenges of certification include conflict over industry standards, as well as consumer confusion due to false claims or multiple certification programs. The ISEAL alliance unites some of the leaders in ethical certification. The alliance represents $53 billion of certified products, which deliver environmental and social benefits to 117 million hectares of agricultural land, and workers at 15,000 factories, fisheries, and farms around the world. Some nonprofit organizations that are certifying bodies include:

Forest Stewardship Council (FSC)
The Forest Stewardship Council (FSC) certifies sustainably harvested wood around the world. It is one of the most successful accreditation programs. Founded in 1993, the FSC sets forth principles, criteria, and standards that span economic, social, and environmental concerns. The FSC system includes stakeholders with a diverse array of perspectives on what represents a well-managed and sustainable forest. FSC standards for forest management have now been applied in over 57 countries around the world. The seventh edition of Harry Potter was printed on FSC-certified paper.

Marine Stewardship Council (MSC)
The Marine Stewardship Council (MSC) is currently the only international standard that assesses whether a fishery is sustainable and well-managed. Assessment is guided by three main principles: the condition of the fish stocks, the impact of the fishery on the environment, and the fishery management in place. Of the 4 million tons of the world's wild fish catch about 7 percent take part in the MSC certification program, and more than 600 fish products in more than 25 countries carry the MSC label.

Marine Aquarium Council
The Marine Aquarium Council sets standards for the collection of marine aquarium fish, including how fish are collected and which fish are collected.

EPEAT—Electronic Product Environmental Assessment Tool

The Electronic Product Environmental Assessment Tool (or EPEAT) was designed to help people select desktop computers, laptops, and monitors based on their environmental attributes. EPEAT also sets out performance criteria for product design and gives manufacturers recognition for their efforts to reduce their product's environmental impact. Products are given a bronze, silver, or gold rate based on the number of criteria they meet. See also http://www.epeat.net.

TRAVEL RESPONSIBLY

Green tourism is difficult to define. Does it only apply to a locally operated jungle lodge in the Amazon, where all the materials are sourced environmentally, or does green tourism extend broadly to include more traditional hotels and resorts that implement some environmental activities, such as reducing water use or amount of linens washed? Arguably, both types are a "shade" of green travel. The unifying theme: travel that reduces your impact on the environment.

The concept of "ecotourism" has evolved since the word was first coined in the 1980s and was the beginning of the "green" travel movement. The exact meaning often varies depending on if it is defined based on the tourist's motivations or from the tourism industry perspective. Ecotourism usually applies to tourism where the primary motivation to travel includes seeing pristine landscapes, and wild animals. However, other segments of the travel industry like "adventure travel" to "cultural tourism" can also be sustainable, green tourism even if the focus is not nature.

Typically ecotourism operations are developed in a way that is sustainable over the long-term, and that integrates with the environment and the community. Certification is limited in the industry and complex, given it involves many countries. Unfortunately, there are also many operators that wear a "green cloak" but are not in fact sustainable tourism. Nature tourism also has a side benefit as it fosters awareness of biodiversity's importance among local residents. Many communities do not realize that their local area and its biodiversity are attractions until tourists arrive. It offers an alternative livelihood, which, when carefully developed, is less destructive than many alternatives in remote natural areas, such as resource extraction and logging.

To add to the famous phrase, "Leave nothing but your footprints," here are a few tips for green travel:

- Choose "green" accommodation, find out about the place you will be staying at, and ask questions about what they do to reduce their resource use and support the community.

- Conserve electricity and water.
- Explore the local culture and eat local foods.
- Leave what you find.
- Keep a safe distance from animals.
- Learn about where you're going and respect cultural differences.

CONSERVE ENERGY

- Buy EnergyStar labeled appliances.
- Select an energy-efficient water heater and insulate it.
- Hang clothes on a rack instead of drying—they will not wear out as quickly either.
- Turn down the thermostat and turn off the air conditioner.
- Unplug appliances and phone chargers when not in use.

LIFE AFTER OIL

There is mounting concern over the impending energy crisis and the dwindling availability of cheap oil—or what many experts refer to as "peak oil." Peak oil refers to the day that world oil production hits its highest mark and begins to decline. As the amount of oil available decreases and our consumption of it continues to rise—the United States currently consumes 21 million barrels of oil a day—demand will exceed supply and *all* goods and services will become prohibitively expensive. Opinions vary on when "peak oil" will happen—many geologists and energy experts believe the peak already occurred, while others place the event 30 years from now. In the United States, domestic oil production peaked in 1972, spurring a huge spike in the price of oil and sparking intense interest in energy conservation measures.

What might a shortage of this nonrenewable resource mean for the average person? Once the price of oil skyrockets, people won't be able to drive everywhere and may be forced to seek jobs closer to their homes. Jobs themselves may diminish as higher operating costs force businesses to close. Food will become expensive to reflect the high cost of transporting produce thousands of miles and growing it with petroleum-derived fertilizers. Air travel will cease to be an option for most people. It will become difficult to heat homes and pump water (on current systems), which is among the various basic services that we consider essential to life. Many people believe that technology will save us, but unless we start developing this technology now, we won't have the capacity to do it in the absence of oil. The U.S. Department of Energy commissioned a report, which concluded that we would need 10 years to prepare for peak oil. A good sign is that alternative sources of energy, such as solar and wind power, are growing in

Figure 5.5 A wind farm near Tarifa in southern Spain. Harnessing wind is becoming an increasingly popular source of alternative energy (*Perkins*)

popularity (see Figure 5.5). The end of oil will, most likely, hasten the need to return to smaller, more localized economies and traditional ways of living.

_____ ✑ _____

CONSERVE WATER

> When the well is dry, we know the worth of water. (Benjamin Franklin [1706–1790], *Poor Richard's Almanac*, 1746)

Although water is essential to life on Earth, many people who live in developed countries and have cheap, regular access to freshwater take it for granted. Few people understand where their water comes from or the connections between their daily activities and the local watershed. Did you know that the amount of water on Earth has remained constant for billions of years? It has continually been recycled since the planet was formed. Ninety-seven percent of water on Earth is saltwater, 2 percent is frozen in the ice caps, and just *1 percent* of the Earth's water is available for human use. If all the earth's water were put in a gallon jug, the freshwater would only make up a bit more than a tablespoon.

Fast Facts About Water

- Groundwater can stay polluted for several thousand years. (Freshwater Society)
- The human body needs about 2.5 quarts of water a day to survive. In the United States, the average person uses 183 gallons a day for cooking, flushing, washing, and watering their lawns. (USGS)
- Every week an estimated 42,000 people die from diseases related to low-quality drinking water and poor sanitation. Over 90 percent of them occur to children under the age of five (WHO/UNICEF).
- 1.2 Billion—The number of people worldwide who do not have access to clean water. (UN)
- 6.8 Billion—The gallons of water Americans flush down their toilets every day. [http://www.nps.gov/rivers/waterfacts.html]
- You can refill an 8-ounce glass of water approximately 15,000 times for the same cost as a 6-pack of soda pop (http://ct.water.usgs.gov/EDUCATION/waterfacts.htm), whereas in some countries, water is more expensive than oil.

Simple Ways to Conserve Freshwater

Don't let the faucet run while you brush your teeth or wash dishes—that's gallons of potable water just flowing down the drain. Install low flow or oxygenated showerheads, or showerheads with a shut-off valve to easily stop the flow of water—you might be surprised how little you actually need the water to run while showering. The toilet is one of the largest water guzzlers in the house; consider replacing older toilets (pre-1992, which consume 6–7 gallons) with a 1.6-gallon or even 1.28-gallon low-flow model. Some places offer rebates for replacing toilets. An alternative if you rent or can't spend the time or money right now on such a project, place two half-gallon milk jugs (weighed down with rocks or dirt) in the tank; this will reduce the water used per flush by one gallon. Throw garbage into a wastebasket, not the toilet. Fix leaks as even a small drip can waste thousands of gallons of water a year. When you wash your clothes be sure to do a full load and also note that the "Permanent Press" cycle on many clothes washing machines consumes an extra 15 gallons of water. Conserving water will save money on your utility bills.

If you water your lawn or garden, do so immediately after the sun goes down to minimize evaporation. If you have a sprinkler system that works on a timer, get a sensor so that it automatically shuts itself off in the event of rain. Make sure the sprinklers are spraying the grass, not the sidewalk. Sweep walkways instead of hosing off debris. The Freshwater Society advises that you wash your car on the lawn instead of the driveway. Water runs off smooth surfaces (such as pavement) and flows straight to the watershed, dropping chemicals into the nearest body of water, but your lawn can act as a filter to break down some of these substances. (Don't forget to shut the hose off while you scrub down the car.)

Filthy water cannot be washed. (West African Proverb)

Opt for lower-phosphate cleaners and detergents. Use a plunger or choose nontoxic cleaners to unclog drains. Don't flush unused medications down the toilet. Painkillers and antibiotics have shown up in drinking water, while hormones in the water supply are believed to be detrimental to the reproductive cycles of fish and other wildlife. Use compost on your lawn instead of phosphorus-laden fertilizers. Most lawns have enough phosphorus already, so any excess flows into nearby bodies of water and creates algal blooms. Use an alternative to salt in the winter to melt snow.

One easy step to protect water is to recycle your used motor oil. If you change your own oil, do not toss the used oil into the garbage or pour it into a storm drain where it will end up flowing into rivers or lakes, and eventually make its way to public beaches and the ground water. The American Petroleum Institute estimates that just 1 gallon of used motor oil can contaminate 1 million gallons of water. If you take your car to a service shop for oil changes, it is good to confirm that they recycle their oil, which is typically required by law. Or, for those of you already embracing water conservation and would like to take it one step further, make your own gray water system. Gray water is wastewater created from typical household cleaning—bathwater, dishwater, and laundry water; it can be used to water your houseplants, lawn, or flower garden. Given that on an average, U.S. households use more water outside the home rather than inside, using gray water outdoors could reap massive water savings. Note, however, that water discharged from the toilet or *black water* is most definitely not suitable for these purposes.

Your gray water system can be simple or elaborate. There are several books and how-to guides on the subject. For your system to be successful, it is important to use biodegradable soaps and avoid harsh bleaching agents. Place a bucket in the shower or sink to catch water while you do chores and then dump it into a larger container, such as a plastic garbage can, or connect the drain hose from the washing machine to your large water storage container; use this water for your plants as needed. With a little extra planning, you can install a gray water system directly into your home. A built-in system like this involves extra pipes to drain wastewater from the tub, washing machine, and/or the kitchen sink, into an underground reservoir, and some way of directing the water outdoors (e.g., through a garden hose or pump through some sort of irrigation system). There are many guidelines to follow to ensure that your system works smoothly and poses no health hazards. Educate yourself on design methods and current gray water laws for your area, or call a plumber for consultation on your project.

You can also collect the rainwater that falls on your roof. The average 25 foot by 40 foot home sheds about 600 gallons of water in an hour of moderate rainfall, around 1 inch. This involves diverting the water from the gutters and downspouts into some sort of reservoir. You can either purchase a rain barrel, which will already have a spigot built into it, or fashion a rain barrel from a large plastic trash can and modify the downspout to flow into the barrel rather than onto the ground. Check the Internet for various Web sites that offer detailed instructions and a list of materials needed for harvesting rainwater. Some people have even been able to filter their harvested water for drinking, but most sources will advise against this and to use this water for gardening purposes only. Learn about your watershed at http://www.epa.gov/surf/.

RECONSIDER THE DAILY COMMUTE

The average commute to work in America is around 30 minutes, but some people travel for 1 or even 2 hours to get to their jobs. Cars account for 25 percent of greenhouse gases emitted in this country and contribute significantly to air pollution.

Consider moving closer to your job, or finding a job closer to your home. Find out if carpooling would be a possibility with any of your co-workers. Take public transportation instead of driving, even if it's just one day a week. Depending on the nature of your work, talk to your boss about telecommuting from home some of the time.

For shorter trips, walk or ride a bike. Why not incorporate your exercise routine into your daily life. Walking at a moderate speed, it takes around 20 minutes to go 1 mile. Consolidate your errands instead of making several short trips every day.

If you're in the market for a new car, consider buying a fuel-efficient hybrid. But don't feel that you should trade in your current car for a hybrid to save gas; the amount of resources and energy consumed to produce any new car could outweigh the benefits of the improved gas mileage (unless you own a Hummer). Decide whether you really need a car at all. Car-sharing organizations exist for people who need a car sometimes but not every day. There is a fee to join and you usually pay per mile or per hour. Fees include maintenance, gas, and insurance, and this can be a significantly cheaper option than owning your own car. Think about contributing to a carbon-offsetting organization to neutralize your emissions. There are a variety of such organizations, and they each work by taking your donations (around $5.00 to $8.00 per ton of CO_2 emissions) and contributing to renewable energy projects or planting trees. Visit www.davidsuzuki.org and click on "Go Carbon Neutral" to learn more and to see a list of organizations. You can contribute some of that gas money you'll save by biking to work now!

GREEN UP YOUR HOME AND GARDEN

In addition to the tips mentioned in the sections on saving energy and water, there are other things you can do around your home to conserve resources and minimize your impact on biodiversity.

For carpentry projects, choose "Forest Stewardship Certified" wood. Bamboo is another highly sustainable option, comparable in look and durability. These days it's much easier to find paints that are good for the planet and easy on the lungs. Avoid latex paint, or paint with VOCs (volatile organic compounds); instead opt for milk- or clay-based paints, or for brands that claim to be solvent-free. If there are any chemicals you have that you won't be using anymore—paint, turpentine, fertilizers, or harsh cleaning agents—take them to a hazardous household waste collection site rather than throwing them in your trash or toilet. Never pour cooking oil down your drain. Look for healthier soaps and detergents; they may seem more expensive, but many are highly concentrated and can be diluted and used for a long time. It wasn't until recently that supermarkets sold so many "extra strength" chemical cleaners. Clean with vinegar and baking soda like in the good old days—both are nontoxic, dirt-cheap, and amazingly effective. Choose soy candles over conventional wax candles, which have been shown to denigrate indoor air quality.

Plant water-efficient, native plants in your yard. Use a drip irrigation system to prevent runoff by applying water directly to the plant's root zone. Stop using pesticides. An estimated 60 million birds are poisoned each year in the United States. Earthworms are vital for soil health, but more than 60 percent are killed by pesticide use. Forego the insect zapper light—they kill more beneficial insects than pests.

Although not immediately obvious, the choices you make about pet care can have ramifications for biodiversity. It is highly recommended that you keep cats indoors, both for their safety and to protect birds, small mammals, and reptiles. Please spay/neuter your animals, especially if they tend to roam around outdoors. Misinformed people who think this procedure is cruel have not considered the cruelty inherent in the fates of thousands of unwanted animals born every day, or the impact that homeless pets have on millions of small animals each year. Don't flush cat litter down the toilet; this spreads the parasite that causes toxoplasmosis, which has been shown to be quite harmful to sea otters. Cat litter itself comes from strip-mining; look for less ecologically invasive brands, such as those made from wheat, corn, pine, or newspaper pellets. Clean up pet waste outside so that it doesn't send contaminants into storm drains.

Build Sustainable Communities

What Makes a Building "Green"?

The U.S. standard for green building is known as LEED, for Leadership in Energy and Design. It establishes standards not only for design and construction but also the operation of green buildings. There are LEED standards for all types of buildings from homes to schools to skyscrapers (see Figure 5.6). LEED looks at sustainability by recognizing building performance in five key areas of human and environmental health: sustainable site development, water savings, energy efficiency, materials selection, and indoor environmental quality. Buildings are assessed based on the following criteria:

* site selection and development;
* water and energy use;
* environmentally preferred construction products, finishes, and furnishings;
* waste stream management;
* indoor environmental quality; and
* innovation in sustainable design.

Depending on which criteria the building meets they can obtain one of the following LEED standards: certified (26–32 points), silver (33–38 points), gold (39–51 points), and platinum (52–69 points). For example, the California Academy of Sciences has been transformed into a green building, and hopes to receive a LEED Platinum status when it

Figure 5.6 Roofs planted with vegetation are becoming an increasingly common feature in major cities. Green roofs reduce heating and cooling costs, limit stormwater runoff, filter pollutants, and provide wildlife habitat. Green roofs also minimize the 'urban heat island' effect, or the tendency for cities to be several degrees warmer because concrete and other building materials absorb and re-emit the sun's heat. This green roof is from downtown Toronto, Canada (*Laverty*)

is completed in 2008. The building includes solar panels, radiant floor heating, uses reclaimedwater, and has a dramatic living roof. The roof is covered with 1.7 million native plants.

Green Buildings Are Going Mainstream

Companies like Walmart are now "greening" their stores across the country. In New York, the new standards are taking the city by storm, with several major skyscrapers built in the last few years being LEED-certified. The Hearst Tower at Eighth Avenue and 57th became the city's first gold standard building when it opened in October 2006; the building is made of 80 percent recycled steel, rain from the roof is collected and stored in the basement where it is used to water plants in the building, and the floor is made of heat conductive limestone. New York's Bank of America Tower will be the first skyscraper rated LEED Platinum: it captures and reuses rainwater; uses recycled material throughout for building; maximizes use of natural light; and generates 70 percent of its own energy. While still a small market, there has been a surge in green building (as well as retrofitting or converting old office buildings to meet green standards) as they are easier to rent. Green workplaces have lower overheads and also create a better and more productive work environment, so companies are taking notice. One myth often connected with green buildings is that the construction costs are substantially higher. In fact in some cases costs can be lower, and even when they are higher the additional costs are typically recovered very quickly due to savings from heat costs.

Empower Individuals and Communities

Some people feel that governments and corporations have more influence on local and global biodiversity, rather than individuals. Some of the most critical priority setting is done everyday by individuals in their own lives—decisions of where to live, what to buy, what to do on and with their land, or even how to vote.

Governments and corporations need our support. In democratic societies, at least, individuals have the responsibility to understand the ramifications of their choices on biodiversity along with the responsibility to participate in local decision-making. Polls show that while 70 percent of U.S. citizens are concerned with the environment, exit polls record that only 28–29 percent of voters actually consider the environment when voting (Dowrie 2001, 2002).

The actions of individuals, whether acting alone or in concert with others, can make a difference to the future of biodiversity. Few individuals truly realize the impact of their daily decisions. In the book *Stuff, the Secret Life of Everyday Things*, authors John Ryan and Alan Durning trace all the environmental costs included in drinking a daily cup of coffee. The decision to drink coffee at all, what brand to purchase, how that coffee was grown and harvested, shipped and distributed, packaged and prepared, all come under consideration. Because many of the steps in coffee production occur elsewhere in the world, there are global ramifications to the decision to drink a simple cup of coffee.

―――――――――――――― ✑∕✑∖ ――――――――――――――

CHAD PREGRACKE

Never doubt that a small group of thoughtful, committed citizens can change the world. Indeed, it is the only thing that ever has. (Margaret Mead, 1901–1978)

Chad Pregracke's inspiring story illustrates Margaret Mead's famous quote and the power of one citizen's actions. Chad grew up near the Mississippi River and worked as a commercial fisherman on the river after graduating from high school. Troubled by what he saw as massive amounts of garbage cluttering the shore, he set off on his own with a small boat to collect (and often recycle) the debris found in and along his "backyard." Others soon took note of his efforts and volunteered to help, and in 1998, he formed his own nonprofit—Living Lands & Waters (LL & W), based in East Moline, Illinois. Today they have a dedicated staff, leagues of volunteers, and a fleet of barges to aid in their ongoing river initiatives. In 2005, they cancelled their fall events to send volunteers and supplies to New Orleans where they helped those in need after Hurricane Katrina. Chad has been recognized internationally and is the recipient of numerous awards for his dedication. As of 2006, LL&W has collected: 15,991 tires, seven port-a-potties, and 30 messages in a bottle. Visit their website (http://www.livinglandsandwaters.org) to see the full list of the types and amounts of garbage collected.

───────────────────────── ∽◦∾ ─────────────────────────

Even when actively supporting conservation, we, as individuals are making decisions that affect biodiversity. Many people support biodiversity by contributing to conservation organizations, yet many of these organizations focus mainly on charismatic megafauna, for example, large furry mammals. This could be at the expense of ecosystem-level action or support for less charismatic or less well-known groups upon which the charismatic fauna depend.

Ultimately, each of the decisions people make, whether consciously or not, is based on what individuals' value and these are the values that will be learned by their children. We must remember that biodiversity is vital to human survival. It is essential for the future of life on the planet that we realize this value and share this knowledge with our friends and family.

GLOSSARY

Albedo: the amount of solar radiation reflected by a surface.

Allopatric speciation: speciation achieved between populations that are geographically completely separated (their ranges do not overlap or are not contiguous).

Anthropocentrism: the notion that humans are the center of the universe and that nature exists (and is used) for human benefit.

Area-sensitive species: species that require large areas of continuous habitat and cannot survive in small patches.

Bequest value: the value of leaving something behind for the next generation.

Bioaccumulation: animals higher up the food chain concentrate contaminants in their bodies.

Biocentrism or Ecocentrism: the notion that life is the center of the universe and humans are a separate but equal part of nature.

Biodiversity: the variety of life on Earth at all its levels, from genes to ecosystems, and the ecological and evolutionary processes that sustain it.

Biodiversity coldspots: areas that have relatively low biological diversity but include rare species or represent unique environments.

Biodiversity hotspots: areas that have high levels of endemism (and hence diversity), but which are also experiencing a high rate of loss of ecosystems. A *terrestrial biodiversity hotspot* is an area that has at least 0.5 percent, or 1,500 of the world's 300,000 species of green plants (Viridiplantae), and that has lost at least 70 percent of its primary vegetation. *Marine biodiversity hotspots* are based on measurements of relative endemism of multiple taxa (species of corals, snails, lobsters, fish) within a region and the relative level of threat to that region.

Biogeography: the study of the distribution of organisms in space and through time.

Biological species concept: a group that interbreeds and is isolated from other groups belonging to the same species

Biome: a major biotic classification characterized by similar vegetation structure and climate, but not necessarily the same species.

Biomimicry: a term to describe research that uses models or takes inspiration from the natural world to solve problems in agriculture, medicine, manufacturing, and commerce.

Biophilia: refers to the notion that the love of nature may have been ingrained into our genes by natural selection.

Brood parasite: a species that places its' young in the nest of another species rather than build their own nests.

Captive breeding: programs at zoos or similar wildlife centers to raise wild animals in captivity, sometimes with the intention to reintroduce them into the wild.

Chemosynthesis: a process whereby bacteria convert carbon molecules into organic matter using hydrogen sulphide, hydrogen, or methane, rather than sunlight, as an energy source

Commercial extinction: when populations are too depleted, or scattered, to be harvested economically.

Community: the populations of different species that naturally occur and interact in a particular environment.

Community stability and resilience: ability to adapt and respond to changing environmental conditions.

Connectivity: the degree to which patches in a landscape are linked.

Conservation easement: restrictions on how land can be used so that it is preserved

Cultivar: a cultivated plant selected for certain desirable characteristics. It is often used interchangeably with the term "variety." Its counterpart for animals is "breeds."

Demography: the statistical characteristics of the population such as size, density, birth and death rates, distribution, and movement or migration.

Direct use value: refers to products or goods that are consumed directly, such as food or timber.

Dominant species: species that are important due to their sheer numbers in an ecosystem.

Ecological biogeography: the study of the dispersal of organisms (usually individuals or populations), and the mechanisms that influence this dispersal, and the use of this information to explain the spatial distribution patterns of the organisms. The focus is on current populations and shorter time scales than historical biogeography.

Ecological extinction: populations may still be present in low numbers, but no longer play important functional roles in the ecosystem.

Ecological footprint: way to measure the demand that people put on the Earth's resources. It can be calculated for an individual, an institution, a community, a city, or a nation.

Ecoregion: is a large area of land or water (typically spanning millions of acres) with a geographically distinct collection of species and natural communities. Several standard methods of classifying ecoregions have been developed, with landform, climate, altitude, ecological processes, and predominant vegetation being important criteria.

Ecosystem: an assemblage of organisms and the physical environment in which they exchange energy and matter.

Ecosystem loss: the disappearance of an assemblage of organisms and their physical environment, such that it no longer functions.

Endemic species: those species whose distributions are naturally restricted to a limited area

Eukaryotes: organisms whose cells contain a nucleus and membrane-bound organelles, such as mitochondria, which carry out specific functions.

Eutrophication: a long-term increase in excess nutrients into an ecosystem.

Evapotranspiration: is the process whereby water is absorbed from soil by vegetation and then released back into the atmosphere.

Evolutionary significant unit (ESU): is defined as a group of organisms that has undergone significant genetic divergence from other groups of the same species. Identifying ESUs requires natural history information, range and distribution data, and a range of genetic analyses.

Existence value: the value people place on just knowing a species or ecosystem exists, even if they will never see it or use it.

Exotic species: species that live outside their native range. The following terms, nonindigenous, nonnative, alien, adventive, neophytes (for plants), and introduced, are synonymous with the term "exotic" species.

Extinct: a species is extinct when there is no reasonable doubt that the last individual has died.

Extinction: the complete disappearance of a species from Earth.

Faunal relaxation: the selective disappearance of some species and replacement by more common species.

Flagship species: large, more charismatic, and well-known vertebrate species (such as spotted owls, gorillas, tigers, and giant pandas), which are often used to promote conservation.

Fragmentation: the subdivision of a formerly contiguous landscape into smaller units.

Gap analysis: a method of examining the distribution of habitat in a landscape to determine where there are "gaps" in protection of a particular species.

Genetic diversity: refers to any variation in the nucleotides, genes, chromosomes, or whole genomes of organisms.

Genetic drift: a random change in allele frequency in a small breeding population leading to a loss of genetic variation.

Genome: the entire complement of DNA within the cells or organelles of the organism.

Genotype: the genetic constitution of an organism that results from the arrangement of the DNA into genes on chromosomes.

Genotypic variation: the variation that exists between the genetic constitution of different individuals.

Global extinction: no living individuals of the species remain anywhere in the world.

Habitat: there are two common usages of the term "habitat." The first defines habitat as a species' use of the environment, while in the second usage it is as an attribute of the land and refers more broadly to habitat for an assemblage of species.

Habitat degradation: the process where the quality of a species' habitat declines.

Habitat loss: the modification of an organism's environment to the extent that the qualities of the environment no longer support its survival.

Historical biogeography: examines past events in the geological history of the Earth and uses these to explain patterns in the spatial and temporal distributions of organisms (usually species or higher taxonomic ranks).

Inbreeding depression: reduction in reproductive ability and survival rates as a result of breeding among related individuals.

Indirect use value: refers to the services that support ecosystems and products that are consumed (this includes ecosystem functions like nutrient cycling and pollination).

Intrinsic or inherent value: describes the worth of an organism, independent of its value to anyone or anything else.

Invasive species: are species whose populations have expanded dramatically, and out-compete, displace, or extirpate native species, potentially threatening the structure and function of intact ecosystems. They can be exotic or native species.

Keystone species: species that have important ecological roles that are greater than one would expect based on their abundance.

Landscape: a mosaic of heterogeneous land forms, vegetation types, and land uses.

Mass extinction: a period when there is a sudden increase in the rate of extinction, such that the rate at least doubles, and the extinctions include representatives from many different taxonomic groups of plants and animals.

Mesopredator release: as large predators disappear, the population of smaller predators often increases.

Metapopulation: a group of different but interlinked populations, with each different population located in its own, discrete patch of habitat.

Morphological species concept: individuals that look alike and share the same identifying traits and belong to the same species.

Nonpoint source pollutant: enters from many locations or is mobile, such as surface runoff into the coastal zone.

Nonuse or passive value: are those things that we don't use but we would feel a loss if they were to disappear. They usually encompass existence and bequest values.

Parapatric: occupying contiguous but not overlapping ranges (referring to the distribution of two or more populations).

Phenotype: the physical constitution of an organism that results from its genetic constitution (genotype) and the action of the environment on the expression of the genes.

Phenotypic variation: the variation of the physical traits, or phenotypic characters of the organism, such as differences in anatomical, physiological, biochemical, or behavioral characteristics.

Photosynthesis: the process by which plants use the energy of sunlight to convert carbon into sugars.

Phylogenetic diversity: the evolutionary relatedness of the species present in an area.

Phylogenetic species concept: a species is a group of individual organisms that share a common ancestor.

Plate tectonics: the forces acting on the large, mobile pieces (or "plates") of the Earth's lithosphere (the upper part of the mantle and crust of the Earth where the rocks are rigid compared to those deeper below the Earth's surface) and the movement of those "plates."

Point source pollutant: enters at a discrete location and is nonmobile, such as effluent from a sewage treatment plant.

Population: a group of individuals of the same species that share aspects of their demography or genetics more closely with each other than with other groups of individuals of that species.

Potential or option value: refers to the use that something may have in the future that has not yet been recognized.

Primary producer: organisms that convert carbon dioxide into organic matter, usually by photosynthesis.

Prokaryotes: organisms that lack a nucleus or membrane-bound organelles.

Protists: organisms part of the Kingdom Protista, either one-celled, or if multicellular, that lack specialized tissues. They cannot be classified as animals, plants, or fungi.

Species diversity: the number of different species in a particular area (i.e., species richness) weighted by some measure of abundance, such as number of individuals or biomass.

Species evenness: the relative abundance with which each species is represented in an area.

Species richness: the number of different species in a particular area.

Strategic value: the value of genes, species, and ecosystems to conservation biologists to increase support for conservation or for scientific research.

Surface roughness: the average vertical relief and small-scale irregularities of a surface.

Sympatric: occupying the same geographic area (referring to the distribution of two or more populations).

Systematics: the study of the diversity of life through time.

Taxonomy: description, identification, classification, and naming of species.

Trophic level: stage in a food chain or web leading from primary producers (lowest trophic level) through herbivores, to primary and secondary carnivores (highest trophic level).

Utilitarian value (also known as Extrinsic, Instrumental, or Use value): refers to something's value as determined by its use or function.

Watersheds: land areas drained by a river and its tributaries.

Wetlands: areas where water is present at or near the surface of the soil or within the root zone all through the year or for a period of time during the year, and where the vegetation is adapted to these conditions.

BIBLIOGRAPHY

Aldrich, R., and J. Wyerman. 2006. *2005 National Land Trust Census Report*. Washington, DC: Land Trust Alliance.

Alverson, D. L., M. Freberg, J. Pope, and S. Murawski. 1994. *A Global Assessment of Fisheries By-Catch and Discards: A Summary Overview*, 1–233. Rome, Italy: FAO Fisheries Technical Paper 339.

Avise, J. C. 1994. *Molecular Markers, Natural History, and Evolution*. New York: Chapman and Hall.

Bailey, R. G. 1983. Delineation of Ecosystem Regions. *Environmental Management* 7: 365–373.

———. 1998. *Ecoregions: The Ecosystem Geography of the Oceans and Continents*. New York: Springer-Verlag.

———. 1998. Ecoregions: Map of North America. Miscellaneous Publication 1548. U.S. Department of Agriculture (USDA), Forest Service. Map 1: 15,000,000.

Barrett, C. B., K. Brandon, C. Gibson, and H. Gjertsen. 2001. Conserving Tropical Biodiversity Amid Weak Institutions. *BioScience* 51(6): 497–502.

Beattie, A., and P. R. Ehrlich. 2004. *Wild Solutions: How Biodiversity Is Money in the Bank*. Second Edition. New Haven, CT: Yale University Press.

Bennett, E. L., and J. G. Robinson. 2000. Hunting of Wildlife in Tropical Forests: Implications for Biodiversity and Forest Peoples. Washington, DC: The World Bank. Available from: http://www-wds.worldbank.org/servlet/WDSServlet?pcont=details&eid=000094946_0103100530377 (accessed on March 1, 2008).

Benyus, J. 2002. *Biomimicry: Innovation Inspired By Nature*. New York: Harper Perennial.

Boesch, Donald, F., H. Richard, Joel E. Burroughs, Robert Baker, P. Mason, Christopher L. Rowe, and Ronald L. Siefert. 2002. *Marine Pollution in the United States*. Washington, DC: Pew Oceans Commission.

Brooks, T. M., G. A. B. Fonseca, and A. S. L. Rodrigues. 2004. Coverage Provided By the Protected-Area System: Is It Enough? *BioScience* 54: 1081–1091.

Bryant, D., D. Nielsen, and L. Tangley. 1997. *The Last Frontier Forests: Ecosystems and Economies on the Edge*. Washington, D.C.: WRI Publications.

Buchmann, S. L., and G. P. Nabhan. 1995. *The Forgotten Pollinators*. Washington, DC: Island Press.

Butler, D. R. 1995. *Zoogeomorphology*. Cambridge, UK: Cambridge University Press.

Carey, C., N. Dudley, and S. Stolton. 2000. *Squandering Paradise? The Importance and Vulnerability of the World's Protected Areas.* Gland, Switzerland: World Wide Fund for Nature International.

Carson, R. L. 1962. *Silent Spring.* Boston: Houghton Mifflin.

Catling, P. M. 2001. Extinction and the Importance of History and Dependence in Conservation. *Biodiversity* 2(3): 1–13. Available from: http://www.tc-biodiversity. org/sample-extinction.pdf (accessed on March 1, 2008).

Center for Biodiversity and Conservation. 1999. *Humans and Other Catastrophes: Perspectives on Extinction.* New York: Center for Biodiversity and Conservation, American Museum of Natural History.

Chape, S., J. Harrison, M. Spalding, and I. Lysenko. 2005. Measuring the Extent and Effectiveness of Protected Areas As an Indicator for Meeting Global Biodiversity Targets. *Philosophical Transactions of the Royal Society of London Biological Sciences* 360(1454): 443–455.

Chevalier, J., J. Cracraft, F. Grifo, and C. Meine. 1997. *Biodiversity, Science and the Human Prospect.* New York: Center for Biodiversity and Conservation, American Museum of Natural History.

Cincotta, R. P., and R. Engelman. 2000. *Nature's Place: Human Population and the Future of Biological Diversity.* Washington, DC: Population Action International.

Coe, J. M., and D. B. Rodgers, editors. 1997. *Marine Debris: Sources, Impacts and Solutions.* New York: Springer-Verlag.

Collar, N. J., L. P. Gonzaga, N. Krabbe, A. Madroño Nieto, L. G. Naranjo, T. A. Parker, and D. C. Wege. 1992. *Threatened Birds of the Americas.* The ICBP/IUCN Red Data Book. Part 2. Third edition. Cambridge, UK: ICBP.

Commission on Geosciences, Environment, and Resources (CGER). 1999. Chapter 4: Marine-Derived Pharmaceuticals and Related Bioactive Agents, 71–82. In *From Monsoons to Microbes: Understanding the Ocean's Role in Human Health. Part II The Value of Marine Biodiversity to Biomedicine.* Available from: http://books.nap. edu/books/0309065690/html/index.html (accessed on March 1, 2008).

Commission on Life Science (CLS). 1999. *Perspectives on Biodiversity: Valuing Its Role in an Everchanging World.* Available from: http://books.nap.edu/books/ 030906581X/html/46.html#pagetop (accessed on March 1, 2008).

Conservation International. 2008. Biodiversity Hotspots. Available from: http:// www.biodiversityhotspots.org/xp/Hotspots (accessed on March 18, 2008).

———. Available from http://www.conservation.org. (accessed on July 11, 2008)

Costanza, R., R. d'Arge, R. de Groot, S. Farber, M. Grasso, B. Hannon, K. Limburg, S. Naeem, R. V. O'Neill, J. Paruelo, R. G. Raskin, P. Sutton, and M. van den Belt. 1997. The Value of the World's Ecosystem Services and Natural Capital. *Nature* 387: 253–260.

Czech, B. and P. R. Krausman 1997. Implications of an Ecosystem Management Literature Review. *Wildlife Society Bulletin* 25(3): 667–675.

Daily, G. C., editor. 1997. *Natures' Services: Societal Dependence on Natural Ecosystems.* Washington, DC: Island Press.

Darwin, Charles. 2006 [1866]. *On the Origin of Species by Means of Natural Selection.* London: John Murray.

DeBeer, J. H., and M. J. McDermott. 1996. *The Economic Value of Non-Timber Forest Products in Southeast Asia, World Conservation Union.* Amsterdam, Netherlands: IUCN.

Dellasalla, D. A., N. L. Staus, J. R. Strittholt, A. Hackman, and A. Iacobelli. 2001. An Updated Protected Areas Database for the United States and Canada. *Natural Areas Journal* 21: 124–135.

DeLong, D. C., Jr. 1996. Defining Biodiversity. *Wildlife Society Bulletin* 24: 738–749.

Devall, B., and G. Sessions. 2001. *Deep Ecology*. Salt Lake City, UT: Gibbs Smith Publisher.

Dowrie, M., and P. Shabecoff. 2001, Spring. Organizing for the Future: An Interview with Mark Dowrie and Philip Shabecoff. *Orion Afield* 6(2): 1.

Dudley, N., and S. Stolton. 1996. *Air Pollution and Biodiversity: A Review*. Gland, Switzerland: World Wildlife Fund International.

Elton, C. S. 1958. *The Ecology of Invasions By Animals and Plants*. London, UK: Methuen & Co.

Erlich, P. R., and A. H. Erlich. 1992. The Value of Biodiversity. *Ambio* 21: 219–226.

Falk, J. H., E. M. Reinhard, C. L. Vernon, K. Bronnenkant, N. L. Deans, and J. E. Heimlich. 2007. *Why Zoos and Aquariums Matter: Assessing the Impact of a Visit to a Zoo or Aquarium*. Silver Spring, MD: Association of Zoos and Aquariums (AZA).

Filardi, C. E., and R. G. Moyle, 2005. Single Origin of a Pan-Pacific Bird Group and Upstream Colonization of Australasia. *Nature* 438: 216–219.

Food and Agriculture Organization of the United Nations (FAO). 2007. State of the World's Forests 2007. Rome, Italy: Food and Agriculture Organization of the United Nations. Available from: http://www.fao.org/docrep/009/a0773e/a0773e00.htm (accessed on May 25, 2007).

———. 2006. The State of World Fisheries and Aquaculture. Rome, Italy. Available from: http://www.fao.org/docrep/009/A0699e/A0699e00.htm (accessed on June 12, 2007).

Franklin, B. 1746. *Poor Richard's Almanack*. Philadelphia, PA: Printed and sold by B. Franklin.

Gaston, K. J., editor. 1996. *Biodiversity: Biology of Numbers and Difference*. Oxford, U.K: Blackwell Science Ltd.

Gaston, K. J. and J. I. Spicer, editors. 2004. *Biodiversity: An Introduction*. Oxford, UK: Blackwell Science Ltd.

Gibbs, J. P., M. L. Hunter, and E. J. Sterling. 2008. *Problem-Solving in Conservation Biology and Wildlife Management: Exercises for Class, Field and Laboratory*. Malden, MA: Blackwell Science. Second edition.

Green, M. J. B., M. G. Murray, G. C. Bunting, and J. R. Paine. 1997. *Priorities for Biodiversity Conservation in the Tropics*. WCMC Biodiversity Bulletin. No. 1.20 pp. Cambridge, UK: World Conservation Monitoring Centre.

Grifo, F., and J. Rosenthal, editors. 1997. *Biodiversity and Human Health*. Washington, DC: Island Press.

Groom, M. J., G. K. Meffe, and C. R. Carroll. 2005. *Principles of Conservation Biology*. Third edition. Sunderland, MA: Sinauer Associates.

Hardin, G. 1968. The Tragedy of the Commons. *Science* 162: 1243–1248.

Hawken, Paul. 2007. *Blessed Unrest: How the Largest Movement in the World Came into Being and Why No One Saw It Coming*. New York: Viking Press.

Hawken, Paul. 1993. *The Ecology of Commerce: A Declaration of Sustainability*. New York: Harper Collins Publishers

Heiser, C. B. 1990. *Seed to Civilization: The Story of Food*. Cambridge, MA: Harvard University Press.

Hoekstra, J. M., T. M. Boucher, T. H. Ricketts, and C. Roberts. 2005. Confronting a Biome Crisis: Global Disparities of Habitat Loss and Protection. *Ecology Letters* 8(1): 23–29.

Hunter, M. Jr., and J. P. Gibbs. 2006. *Fundamentals of Conservation Biology*. Third edition. Malden, MA: Blackwell Science.

Hutchinson, G. E. 1959. Homage to Santa Rosalia or Why Are There So Many Kinds of Animals. *The American Naturalist* 93(870): 145–159.

Intergovernmental Panel on Climate Change (IPCC). 2007. *Climate Change 2007: Synthesis Report. Contribution of Working Groups I, II, III to the Fourth Assessment Report of the Intergovernmental Panel on Climate Change.* [Core Writing Team: R. K. Pachauri and A. Reisinger, editors.] Geneva, Switzerland: IPCC.

International Human Genome Sequencing Consortium. 2001. Initial Sequencing and Analysis of the Human Genome. *Nature* 409: 860–921.

[IUCN] World Conservation Union. 2008. IUCN Red List of Threatened Species. Available from: http://www.redlist.org (accessed on March 11, 2008).

Jackson, J. B. C. 2001. *What Was Natural in the Coastal Ocean? Proceedings of the National Academy of Sciences* 98(10): 5411–5418.

Kellert, S. R., and E. O. Wilson. 1993. *The Biophilia Hypothesis*, Washington, DC: Island Press.

Laurence, W. F., and R. O. Bierregaard, Jr., editors. 1997. *Tropical Forest Remnants*, Chicago, IL: University of Chicago Press.

Lawton, J. H., and R. M. May, editors. 1995. *Extinction Rates*. Oxford, UK: Oxford University Press.

Lecointre, G. and H. Le Guyader. 2001. *Classification phylogenetique du vivant*. Paris: Berlin.

Leopold, Aldo. 1993. *Round River*. Luna Leopold, editor. New York: Oxford University Press, 145–146.

Levin, P. S., and D. A. Levin. January-February, 2002. The Real Biodiversity Crisis. *American Scientist* 90(1): 6–8.

Linnaeus, Carolus. 1735. *Systema Naturae*. Leiden, Netherlands: Apud Theodore Haak.

Lomolino, M. V., B. R. Riddle, and J. H. Brown. 2006. *Biogeography*. Third edition. Sunderland, MA: Sinauer Associates.

Lynn, W. S. 2002. Canis Lupus Cosmopolis: Wolves in a Cosmopolitan Worldview. *Worldviews* 6(3): 300–327.

Mack R. N., D. Simberloff, W. M. Lonsdale, H. Evans, M. N. Clout, and F. Bazzazz. 2000. Biotic Invasions: Causes, Epidemiology, Global Consequences and Control. *Issues in Ecology* 5: 1–20.

Maddison, D. R. 2008. The Tree of Life Web Project. Available from: http://beta.tolweb.org/tree (accessed on March 20, 2008).

Magurran, A. E. 2003. *Measuring Biological Diversity*. Malden, MA: Blackwell Science.

Malcolm, J. R., and A. Markham. 2000. *Global Warming and Terrestrial Biodiversity Decline*. Washington, DC: World Wildlife Fund.

Malcolm, J. R., and L. F. Pitelka. December 2000. *Ecosystems and Global Climate Change: A Review of Potential Impacts on U.S. Terrestrial Ecosystems and Biodiversity*. Prepared for the Pew Center on Global Climate Change.

Marsh, G. P. 1864. *Man and Nature: The Earth as Modified by Human Action*. New York: Charles Scribner.

Matthews, E., R. Payne, M. Rohweder, and S. Murray. 2000. Pilot Analysis of Global Ecosystems (PAGE): Forest Ecosystems. World Resources Institute. Available from http://www.wri.org/publication/pilot-analysis-global-ecosystems-forest-ecosystems (accessed on March 17, 2008).

Mayr, E., and P. D. Ashlock. 1991. *Principles of Systematic Zoology*. New York: McGraw-Hill.

McDonough, W., and M. Braungart. 2002 *Cradle to Cradle: Remaking the Way We Make Things*. New York: North Point Press.

McNab, W. H., and P. E. Avers. 1994. Ecological Subregions of the United States. WO-WSA-5. Prepared in cooperation with Regional Compilers and the ECOMAP Team of the Forest Service. Available from: http://www.fs.fed.us/land/pubs/ecoregions/index.html (accessed on March 21, 2008).

Meffe, G. K., L. A. Nielsen, R. L. Knight, and D. A. Schenborn. 2002. *Ecosystem Management: Adaptive, Community-Based Conservation*. Washington, DC: Island Press.

Mooney, H. A., and R. J. Hobbs. 2000. *Invasive Species in a Changing World*. Washington, DC: Island Press.

Moran, D., and D. Pearce. 1994. *The Economic Value of Biodiversity, World Conservation Union*. London, U.K: Biodiversity Programme, Earthscan Publications.

Murray, John. 1895. *Report of the Scientific Results of the Voyage of H.M.S. Challenger*. Editorial Notes. London: Eyre and Spottiswoode.

Myers, N., and J. Kent. 2001. *Perverse Subsidies: How Misused Tax Dollars Harm the Environment and the Economy*. Washington, DC: Island Press.

Myers, N., R. A. Mittermeier, C. G. Mittermeier, G. A. B. da Fonseca, and J. Kent. 2000. Biodiversity Hotspots for Conservation Priorities. *Nature* 403: 853–858.

The Nature Conservancy. Available from: http://www.tnc.org. (accessed on July 11, 2008).

Newmark, W. D. 2002. *Conserving Biodiversity in East African Forests: A Study of the Eastern Arc Forests*. New York: Springer-Verlag.

Noss, R. F. 1990. Indicators for Monitoring Biodiversity: A Hierarchical Approach. *Conservation Biology* 4: 355–364.

Novacek, M. J. 2007. *Terra: Our 100-Million-Year-Old Ecosystem and the Threats That Now Put It At Risk*. New York: Farrar, Straus and Giroux.

———. 2000. *The Biodiversity Crisis: Losing What Counts*. New York: The New Press.

Oerke, E. C., H. W. Dehne, F. Schonbeck, and A. Weber, 1994. *Crop Production and Crop Protection*. Amsterdam, The Netherlands: Elsevier Science.

Ohrnberger, D. 1999. *The Bamboos of the World*. Amsterdam. The Netherlands: Elsevier Science.

Oldfield, M. L. 1984. *The Value of Conserving Genetic Resources*. U.S. Department of the Interior. Washington, DC: National Park Service.

Olson, D. M., and E. Dinerstein. 2002. The Global 200: Priority Ecoregions for Global Conservation. *Annals of the Missouri Botanical Garden* 89: 199–224.

Oosterzee, Penny. 1997. *Where World's Collide: The Wallace Line*. Ithaca, NY: Comstock Books.

Parkinson, C. L. 1997. *Earth from Above: Using Color-Coded Satellite Images to Examine the Global Environment*. Sausalito, CA: University Science Books.

Pauly, D., V. Christensen, J. Dalsgaard, R. Froese, and F. Torres, Jr. 1998. Fishing Down Marine Food Webs. *Science* 279: 860–863.

Perlman, D. L., and G. Adelson. 1997. *Biodiversity: Exploring Values and Priorities in Conservation.* Malden, MA: Blackwell Science.

Pimentel, D., L. Lach, R. Zuniga, and D. Morrison, 2000. Environmental and Economic Costs Associated with Non-Indigenous Species in the United States. *Bioscience* 5(1): 53–65.

Pimentel, D., C. Harvey, P. Resosudarmo, K. Sinclair, D. Kurz, M. McNair, S. Crist, L. Shpritz, L. Fitton, R. Saffouri, and R. Blair. 1995. Environmental and Economic Costs Of Soil Erosion And Conservation Benefits. *Science* 267: 1117–1123

Primack, R. B. 2006. *Essentials of Conservation Biology.* Fourth edition. Sunderland, MA: Sinauer Associates.

Raup, D. M. 1991. *Extinction: Bad Genes Or Bad Luck?* New York: W. W. Norton and Co.

Reaser, J. K. 2000. *Amphibian Declines: An Issue Overview.* Washington, DC: Federal Task Force on Amphibian Declines and Deformities.

Redford, K. H., 1992. The Empty Forest. *BioScience* 42: 412–422.

Riebel, L., and K. Jacobsen. 2002. *Eating to Save the Earth.* Berkeley, CA: Celestial Arts.

Roberts, C. M., C. J. McClean, J. E. N. Veron, J. P. Hawkins, G. R. Allen, D. E. McAllister, C. G. Mittermeier, F. W. Schueler, M. Spalding, F. Wells, C. Vynne, and T. B. Werner. 2002. Marine Biodiversity Hotspots and Conservation Priorities for Tropical Reefs. *Science* 295: 1280–1284.

Robinson, J. G., and E. L. Bennett, 2000. *Hunting for Sustainability in Tropical Forests.* New York: Columbia University Press.

Rogers, E., and T. M. Kostigen. 2007. *The Green Book: The Everyday Guide to Saving the Planet. One Simple Step at a Time.* New York: Three Rivers Press.

Roe, D. T. Mulliken, S. Milledge, J. Mremi, S. Mosha, and M. Grieg-Gran. 2002. *Making a Killing Or Making a Living? Wildlife Trade, Trade Controls and Rural Livelihoods. Biodiversity and Livelihoods Issues No. 6. TRAFFIC.* International Institute of Environment and Development. Herts, UK: Earthprint, Stevenage.

Roper, J., and R. W. Roberts. January 1999. Deforestation: Tropical Forests in Decline. Quebec, Canada: Canadian International Development Agency.

Ryan, J., and A. T. Durning. 1997. *Stuff: The Secret Lives of Everyday Things.* Seattle, CA: Northwest Environment.

Sagoff, M. 1988. *The Economy of the Earth.* New York: Cambridge University Press.

Salati, E. 1987. The Forest and the Hydrological Cycle, 273–294 In R. Dickinson, editor. *The Geophysiology of Amazonia.* New York: John Wiley and Sons.

Sanderson, E. W., M. Jaiteh, M. A. Levy, K. H. Redford, A.V. Wannebo, and G. Woolmer. 2002. The Human Footprint and the Last of the Wild. *BioScience* 52(10): 891–904.

Schopf, J. W., editor. 1983. *Earth's Earliest Biosphere. Its Origin and Evolution.* Princeton, NJ: Princeton University Press.

Scott, J. M., R. J. F. Abbitt, and C. R. Groves. 2001. What Are We Protecting? The U.S. Conservation Portfolio. *Conservation Biology in Practice* 2(1): 18–19.

Sechrest, W., T. M. Brooks, G. A. B. da Fonseca, W. R. Konstant, R. A. Mittermeier, A. Purvis, A. B. Rylands, and J. L. Gittleman. 2002. Hotspots and the Conservation of Evolutionary History. *Proceedings of the National Academy of Sciences* 99(4): 2067–2071.

Serafy, S. E. 1998. Pricing the Invaluable: The Value of the World's Ecosystem Services and Natural Capital. *Ecological Economics* 25(1): 25–27.

Soulé, M. E., and M. A. Sanjayan. 1998. Conservation Targets: Do They Help? *Science* 279: 2060–2061.

Speer, B. R., and A. G. Collins. 2000. *University of California Museum of Paleontology Taxon Lift.* B. R. Speer and A.G. Collins, editors. Available from: http://www.ucmp.berkeley.edu/help/taxaform.html (accessed on March 20, 2008).

Stein, B. A., L. S. Kutner, and J. S. Adams. 2000. *Precious Heritage: The Status of Biodiversity in the United States.* Oxford, UK: Oxford University Press.

Tilman, David. 1999. The Ecological Consequences of Changes in Biodiversity: A Search for General Principles. *Ecology* 80(5): 1455–1474.

Toman, M. 1998. Why Not to Calculate the World's Ecosystem Services and Natural Capital. *Ecological Economics* 25(1): 57–60.

Tudge, C. 2000. *The Variety of Life.* Oxford, UK: Oxford University Press.

Turner, R. K., W. N. Adger, and R. Brouwer. 1998. Ecosystem Services Value, Research Needs, and Policy Relevance: A Commentary. *Ecological Economics* 25(1): 61–65.

Ulrich, R. S. 1986. Human Responses to Vegetation and Landscapes. *Landscape and Urban Planning.* 13: 29–34.

United Nations Millennium Ecosystem Assessment. Available from: http://www.millenniumassessment.org/en/index.aspx (accessed on March 20, 2008).

United States Department of Agriculture (USDA). 2007. Cotton Briefing Room. USDA Economic Research Service. Available from: http://www.ers.usda.gov/Briefing/Cotton/ (accessed on March 20, 2008).

Vandermeer, J., I. Perfecto, and V. Shiva. 2005. *Breakfast of Biodiversity: The Political Ecology of Rainforest Destruction,* 2nd ed. Oakland, CA: Food First.

Van DeVeer, D., and C. Pierce.1998. *The Environmental Ethics and Policy Book.* Second edition. New York: Wadsworth Publishing Company.

Van Dyke, F. 2003. *Conservation Biology: Foundations, Concepts, Applications.* New York: McGraw Hill.

Veron. J. 2000. *Corals of the World.* Townsville, MC, Queensland, Australia: Australian Institute of Marine Science and CRR Qld Pty Ltd.

Vietmeyer, N. 1996. New Crops: Solutions for Global Problems, 2–8. In J. Janik, editor, *Progress in New Crops.* Alexandria, VA: American Society of Horticultural Science Press.

Vitousek, P. M., H. A. Mooney, J. Lubchenco, and J. M. Meiillo. 1997. Human Domination of Earth's Ecosystems. *Science* 277: 494–499.

Wilson, E. O. 1999. *The Diversity of Life.* New York: W.W. Norton.

World Charter for Nature. 1982. United Nations General Assembly Resolution A/RES/37/7 Available from: http://www.un.org/documents/ga/res/37/a37r007.htm (accessed on March 1, 2008).

World Commission on Dams. November 2000. *Dams and Development: A New Framework for Decision-Making.* The Report of the World Commission on Dams. London, UK: Earthscan Publications.

World Conservation Monitoring Centre, United Nations Environment Programme. Available from: http://www.unep-wcmc.org (accessed on July 11, 2008).

World Conservation Union (IUCN). Available from: http://www.iucn.org (accessed on July 11, 2008).

World Wide Fund for Nature (WWF). WWF Conservation Science Program: Ecore-
    gions. Available from: http://www.worldwildlife.org/science/ecoregions.cfm
    (accessed on March 19, 2008).

World Wide Fund for Nature (formerly World Wildlife Fund). Available from:
    http://www.wwf.org (accessed on July 11, 2008).

Index

About the Authors

MELINA F. LAVERTY is a marine biologist, and former Senior Program Officer at the American Museum of Natural History's Center for Biodiversity and Conservation. She is currently completing her law degree at the University of Toronto.

ELEANOR J. STERLING is the Director of the Center for Biodiversity and Conservation at the American Museum of Natural History, as well as the Director of Graduate Studies for the Department of Ecology, Evolution, and Environmental Biology at Columbia University.

AMELIA CHILES is a former staff member of the American Museum of Natural History's Center for Biodiversity and Conservation. She now lives in Denver where she works for the Colorado chapter of the U.S. Green Building Council.

GEORGINA CULLMAN works in the Center for Biodiversity and Conservation at the American Museum of Natural History.